RESPONSES OF ORGANISMS TO WATER STRESS

Edited by **Şener Akıncı**

Responses of Organisms to Water Stress
http://dx.doi.org/10.5772/46157
Edited by Şener Akıncı

Contributors

Hamadeh, Allan Lobato, Elaine Guedes, Cândido Oliveira Neto, Douglas Marques, Roberto Costa, Joaquim Siveira, Elizamar Ciríaco Da Silva, Manoel Albuquerque, André Azevedo Neto, Carlos Silva Junior, Alexandre Bosco de Bosco De Oliveira, Nara Lídia Mendes Alencar, Enéas Gomes-Filho, Masaharu Motoshita, Sonia Marli Zingaretti, Marielle Cascaes Inacio, Livia De Matos Pereira, Suzelei De Castro França, Tiago Antunes Paz

Published by InTech

Janeza Trdine 9, 51000 Rijeka, Croatia

Publishing Process Manager Iva Lipovic
Technical Editor InTech DTP team
Cover InTech Design team

Additional hard copies can be obtained from orders@intechopen.com

Responses of Organisms to Water Stress, Edited by Şener Akıncı
p. cm.
ISBN 978-953-51-0933-4

For EU product safety concerns:
IN TECH d.o.o., Prolaz Marije Krucifikse Kozulić 3, 51000 Rijeka, Croatia,
info@intechopen.com or visit our website at intechopen.com.

Meet the editor

Dr. Şener Akıncı received his PhD degree in 1997 from Sheffield University, England. Since then he has been studying Stress Physiology at the Department of Biology, Faculty of Arts and Sciences, Marmara University, Turkey, where he is Vice Director of the " Native Plants and Water Products Applied Research Center". He has published more than 15 original research articles nationally and internationally, and one book chapter in 2012 issued by In Tech. He is currently working as Associate Professor at the Botany section, Department of Biology, Faculty of Arts and Sciences, Marmara University, Turkey.

Contents

Preface

Water is a fundamental requirement for life and an essential factor for all organisms, from cells to whole body, and from first cell division until death. Globally only 2.5% of water is present as fresh water, of which about 68% is in glaciers and 30% in ground water. The rest is to be found as atmospheric humidity, surface water in the form of rivers and lakes, soil moisture, and in plants and animals. Water has a crucial role as a permanent substance of the central vacuole in plant cells, with the water component ranging from 85-95% in fresh leaves and young tissues, 35-75% in woody parts and stems, and 5-15% in dry seeds.

Water stress is one of the major environmental factors that affects most terrestrial organisms, and in plants leads to readily distinguishable effects on growth parameters, accompanied by changes in biomass ratios and physiological and biochemical alterations. Stress symptoms are visible morphologically and as biomass reduction depending on the severity and duration of drought exposure. Water stress (drought) decreases plant water potential and turgor, causing physiological difficulties, inhibition of photosynthesis and respiration, effects on metabolic and biochemical processes, changes in carbohydrate content, quantity and quality of nutrients, translocation, lipid composition in leaves, and plant hormone regulation.

Water stress not only effects plant-animal community interactions but also human societies, as a result of impacts on horticultural systems and agricultural lands, as well as natural ecosystems. Every year many cultivated areas of the world experience drought, particularly in arid and semi-arid climates. Water loss and lack of water availability from soil is therefore of considerable importance in agricultural and horticultural areas, where crop production mostly depends directly on precipitation regimes, since use of irrigation is limited on a world scale. It is well known that drought can cause more than 50% of yield reduction in most crop plants. The United Nations' FAO states that by 2025, 1.9 billion people will be living in countries or regions with absolute water scarcity, and two-thirds of the world population could be under water-stress conditions. Since about one-third of potential arable land is facing water scarcity, and yield production in the remainder may be adversely affected by periodic drought, FAO reports state that more than half of the world population could be negatively affected by 2025.

The editor hopes that this wide-ranging book, with seven chapters, will be beneficial for all those interested in plant-water research, including students, researchers from scientific institutions and universities, and other professionals. The editor cordially extends his thanks to the authors, who are from all over the world, for their valuable contribution to the book. He also would like to express his appreciation particularly to Ms. Daria Nahtigal, Ms. Maria

Jozipovic and Ms. Iva Lipovic from InTech Open Access Publisher for their great effort and support throughout this whole processes of publishing the "water stress" book.

Dr. Şener AKINCI
University of Marmara
Turkey

Chapter 1

Quantification of Stress Arisen from Freshwater Consumption in the Context of Life Cycle Assessment

Masaharu Motoshita

Additional information is available at the end of the chapter

http://dx.doi.org/10.5772/54237

1. Introduction

Freshwater is one of the most essential resources for living things on the earth. Increasing water demand due to population and economic growth in the world may threat the balance of freshwater supply and demand. Consequently, almost 30% of world population is expected to be suffering from water scarcity in 2025 according to the UNESCO's prospects [1]. Physical scarcity of freshwater will cause several kinds of stress on human and ecosystem. In order to avoid or minimize the effects of freshwater scarcity, the balance of freshwater demand and resource amount should be managed appropriately.

Freshwater is consumed not only directly but also indirectly in our activities. For instance, a cup of coffee directly requires freshwater for dripping coffee and washing a cup and drip equipment. In addition, freshwater is indirectly consumed for making a cup of coffee through the life cycle (growing coffee plants, processing coffee beans, producing packaging and so on) [2-3]. Thus, freshwater consumption should be analyzed and managed in the context of life cycle thinking.

As a tool for accounting stress of freshwater consumption based on life cycle concept, water footprinting has attracted high attention in recent years. Water footprinting generally accounts both the volume of consumed freshwater and the impact resulting from freshwater consumption. The stress of freshwater consumption will be different among regions. In this context, to quantify the impact of freshwater consumption with the consideration of regional differences has been seemed to be of significance and several researches on this topic have been performed for modelling the impact of freshwater consumption as life cycle impact assessment model.

The stress arisen from freshwater consumption can be identified in two steps (midpoint and endpoint) in accordance with general life cycle impact assessment methodology. In the midpoint assessment, physical scarcity of freshwater due to consumption is quantified by considering freshwater availability in each region. Endpoint assessment focuses on more concrete damage caused by freshwater consumption. The details of advanced knowledge on quantifying stress of freshwater consumption, from physical scarcity to concrete damage on human and ecosystem, in several researches will be introduced in the following sections as state-of-the-art activities for accounting water stress in the quantitative aspect.

2. Midpoint assessment

The critical problem of water consumption is the availability loss of freshwater for downstream users. If withdrawn freshwater were returned to the original basin without any quality degradation (chemical and thermal), the availability of freshwater for downstream users are not restricted and no stress can be arisen. In such case, the amount of withdrawn water is defined as "water use" and excluded from accounting the stress of freshwater consumption [4-5]. Disappeared and/or degraded amount of freshwater is defined as "water consumption" and accounted for assessing the stress of freshwater consumption in both midpoint and endpoint assessment.

Midpoint assessment in life cycle impact assessment is the step to quantify the scientifically clear and category specific change in the environment. For instance, greenhouse gas emission will cause the change of radiative forcing and result in human health damage like malaria and dengue fever. While human health damage is a common issue among different environmental categories, the change of radiative forcing is a unique natural phenomenon relevant to global warming. Thus, the change of radiative forcing is generally selected as the indicator of global warming at midpoint level. In accordance with this concept of life cycle impact assessment, physical scarcity of freshwater is defined in most researches as the indicator of freshwater consumption stress at midpoint level.

Several methods on midpoint assessment have been proposed [5-10]. The basic and common concept of impact assessment indicator on freshwater consumption at midpoint level is the ratio of consumed amount of freshwater to the amount of available freshwater resources, indicating physical scarcity of freshwater as shown in equation 1.

$$\text{The impact indicator} = \frac{\text{Consumed amount of freshwater}}{\text{The amount of available freshwater}} \quad (1)$$

Methods on midpoint assessment can be characterised by the consideration of influential factors (the threshold of available freshwater resource amount, temporal variation, spatial differences, non-linearity of sensitivity to scarcity and quality of freshwater resources). Characteristics of each method in the above describe five factors are as follows.

1. The threshold of available freshwater resource amount

   All the amount of freshwater resources is not necessarily available. Thus, some methods applied threshold amount of freshwater resources [5-8]. Frischknecht et al. [5] adopted 20% of total freshwater resources as a threshold based on expert judgement. Mila i Canals et al. [6] and Hoekstra et al. [7] considered environmental water requirement including ecosystem as an elementary water demand. The difference between total amount of freshwater resource and environmental water requirement is defined as the amount of available freshwater in their methods. Boulay et al. [8] differentiated surface water from groundwater as freshwater resources and defined 90% low flow (the low flow is exceeded in 9 month out of 10) of surface water as the threshold in order to exclude unusual high flow effects. Determination of a threshold of freshwater resource is different among methods and generally performed by expert judgment, and it can be a critical argument point.

2. Temporal variability

   The amount of freshwater resource tends to have temporal variation (ex. differences between the dry seasons and the rainy seasons). The monthly variation of available freshwater resource (river runoff) was estimated by Hoekstra et al. [7]. Actually, stored freshwater (like pond, lake, dam and so on) can be available freshwater resource in addition to flowing water. Pfister et al. [9] considered temporal variation of precipitation (monthly and annual) in assessing available freshwater resource including stored water by introducing variation factor of annual and monthly precipitation.

3. Regionalized differences

   Freshwater supply by precipitation and influential factors on that (like climate and landform condition) are not even on the earth. Thus, the availability of freshwater is spatially different. Spatial difference is taken into account in each method on different resolution (on country scale to grid scale). Detailed resolution would be preferable in the context of science. However, very detailed site specification might be not necessarily practical because supply chain of products and companies are too complicated to specify the precise location of consumed freshwater. Both of preciseness and applicability should be harmonised from the view point of practical use.

4. Non-linearity of sensitivity to scarcity

   The increase of freshwater consumption results in increasing the impact of physical water scarcity, but obviously the rate of the increase will not be equal between resource abundant and scarce area. In the Swiss Ecological Scarcity Method [5], the ratio of critical water flow and current water flow was squared to reflect the severity in freshwater scarce region and the strength in freshwater abundant region. Pfister et al. [9] described non-linearity between available freshwater resource amount and impact of freshwater consumption by adjusting equation 1 to a logistic function. As a result, resource abundant areas are not sensitive to freshwater availability change, and resource scarce areas are sensitive to that. Potential adaptability to freshwater consumption in the physical aspect of freshwater resources is reflected in the method. On the other hand, Boulay et al. [8] also considered non-linearity between withdrawal-based and consumptive-based amounts of freshwater by applying the S-curve fitting on the basis of regression analysis. This method seems to focus on the adaptability to freshwater consumption in the social aspect of freshwater use rather than physical aspect of resources.

5. Quality of freshwater resources

Freshwater availability will be also controlled by the quality of resources and of emitted/returned water. From the perspective of input freshwater quality, the freshwater availability of downstream user depends on the quality of resource even if the same amount is consumed. Pure quality freshwater can be used by most users but degraded freshwater in chemical/thermal composition will be available for only limited users. "Gray water" is one of the concepts to reflect the impact of quality degradation of water. The emissions with used water will demand freshwater for the dilution of the emissions to avoid restricting downstream users' availability. The amount of freshwater enough to diminish the emissions to the acceptable level (generally environmental criteria of the basin) is regarded to be consumed virtually. Gray water is the amount of assumed freshwater volume for the dilution. This concept was adopted to take the quality degradation into account in two studies [7, 10]. A point to notice is that gray water is not actually consumed freshwater but virtually assumed consumptive freshwater. Boulay et al. [8] developed the impact indicators correspond to the quality of freshwater resource by considering threshold value of the quality for each user's demand. In addition, their method can assess the impact in quality of not only input water but also output water by calculating the difference between negative effect of withdrawn water and positive effect of returned water.

In the context of midpoint assessment, existing methods have unique characteristics by considering a different combination of above aspects. Thus, the relevance of each aspect is difficult to be clarified through simple comparison of impact factors of each method. On the other hand, the consideration of influential factors on the impact of freshwater scarcity made it possible to reflect the actual situation relevant to freshwater scarcity. For instance, rank of renewable freshwater resource per capita in each country [11] and impact factors on freshwater consumption developed by Pfister et al. [9] are shown in Figure 1, Figure 2, respectively. Higher ranked countries (severe to water scarcity) are deeply colored in Figure 1, Figure 2. Severity in resource amount and impact factor shows similarity in some countries but difference in others. A typical difference can be seen in Australia. While the amount of freshwater resource is abundant, stress to water scarcity is relatively higher. Method of Pfister et al. [9] integrated temporal variation of precipitation, and actually draught has sometimes occurred in Australia. Such a real condition in some aspects could be reproduced in existing methods on midpoint assessment. However, it should be verified through the comparison with endpoint assessment model whether a midpoint assessment model is adequate to represent the final consequences of freshwater consumption.

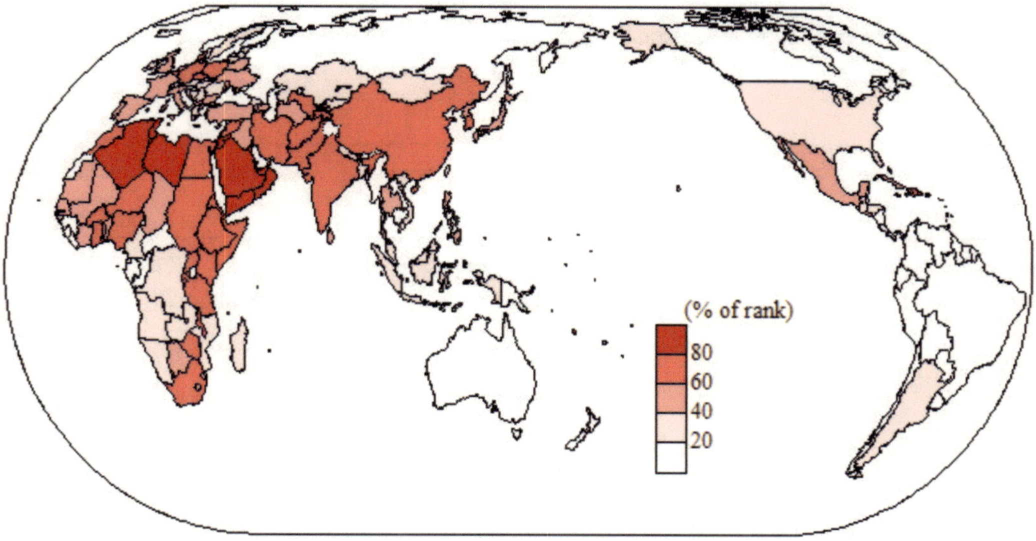

Figure 1. Renewable freshwater resource per capita in each country

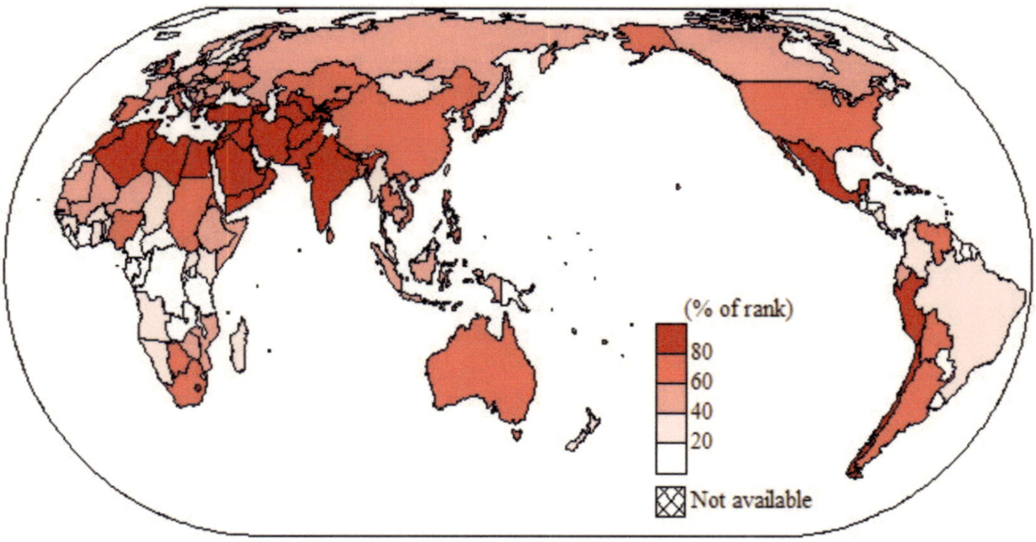

Figure 2. Impact factors (water stress index per unit volume freshwater consumption) of each country [9]

3. Endpoint assessment

Freshwater consumption will cause several kind of damage on human and ecosystem through physical water scarcity. As major endpoints of freshwater consumption, damage on human health, ecosystem and resources is modelled in several studies. Classification of endpoints and corresponding assessment methods are summarized in Table 1. Details of modelling on each endpoint are explained in the following sections.

Users of freshwater suffering from scarcity		Endpoint and specific consequences		Corresponding assessment methods
Human society	Domestic water	Human health	Increasing damage of infectious diseases	Boulay et al. [8], Motoshita et al. [13]
	Agricultural water		Increasing damage of malnutrition	Boulay et al. [8], Pfister et al. [9], Motoshita et al. [15]
		Resources	Agricultural, animal and aquacultural commodity production loss	Motoshita et al. [15]
	Industrial water		Economic production loss	No method available
	All users		Surplus energy demand for compensation	Pfister et al. [9]
Ecosystem	Terrestrial species	Ecosystem	Plant growth prevention	No method available
			Species extinction due to habitat loss	Pfister et al. [9], van Zelm et al. [18]
	Aquatic species			Maendly and Humbert [19]

Table 1. Classification of endpoint relevant to freshwater consumption

3.1. Human health

Human health damage is one of the most major endpoints as a consequence of freshwater consumption. According to the report of World Health Organization (WHO), almost 9% of total health damage (including both mortality and morbidity) in the world is estimated to be arisen from water, sanitation and hygiene [12]. Particularly, diarrhoeal disease and malnutrition are account for over 70% of water-related health damage, and they seemed to be highly related to the availability of freshwater. Thus, human health damage of infectious diseases and malnutrition due to freshwater consumption has been quantified in previous studies [8, 9, 13, 15].

Infectious diseases will be arisen from the intake of low quality water in the context of freshwater consumption. Damage of four infectious diseases (Ascariasis, Trichuriasis, Diarrhoea, Hookworm disease) related to freshwater consumption was modelled by Motoshita et al.

[13]. The relationship between infectious disease damage and freshwater availability loss on country scale was analyzed based on statistical data by applying multiple-regression model with the consideration of social and economic factors (GDP per capita, capital formation expenditure per capita, temperature, accessibility to safe water/sanitation, nutritional condition and medical treatment opportunity). Boulay et al. [8] evaluated damage of both diarrheal disease and nematode infections caused by freshwater consumption in each country. Health damage due to freshwater consumption on country average was estimated by dividing a deficit volume of freshwater (the difference between actual use and minimum requirement of domestic water) into damage of target diseases per country. Country specific social condition was also considered by introducing the adaptation capacity parameter using gross national income (GNI).

The shortage of freshwater for food production as a consequence of freshwater consumption will cause the nutritional deficit. On the other hand, social and economic conditions in each region will control the effects of nutritional deficit due to freshwater consumption. In the method of Pfister et al. [9], Human Development Index (HDI) was adopted as an explanatory indicator for social and economic condition. HDI is an indicator for representing development degree of each country with the consideration of health (average life expectancy), education (adult literacy and gross enrolment) and economic level (gross domestic production per capita) [14]. The relationship between malnutrition damage and HDI was modelled by regression analysis based on statistical data on country scale and was adjusted from 0 to 1 to reflect the vulnerability to nutritional deficit due to freshwater consumption in each country. More straightforward factors were used to explain the relationship between malnutrition and water scarcity in the modelling by Motoshita et al. [15]. Parameters on nutritional and medical conditions (average food consumption level, gaps in food consumption (Gini coefficient) and medical treatment expenditure per capita) were applied to malnutrition damage modelling by using multiple regression analysis. In addition, food shortage in a country will spread to other countries through international trade. Such a ripple effect was also integrated into the modelling to reflect the interaction among countries. While Boulay et al. [8] simply estimated malnutrition damage due to freshwater shortage by dividing the water requirement per calorie into malnutrition damage per unit total calorie deficit on country scale, differences of social and economic situations among countries were considered by applying adaptation capacity parameter (GNI) as same as the modelling on domestic water scarcity. Aquaculture is one of the nutritional resources in some countries. Boulay et al. [8] considered the effect of freshwater shortage in aquaculture while other two methods [9, 15] on malnutrition damage did not consider.

The significance of infectious disease and malnutrition damage can be compared based on the characterisation factors of Motoshita et al. [13, 15]. Both damage of infectious disease and malnutrition caused by freshwater consumption on country scale was shown in Figure 3. Malnutrition damage due to agricultural water scarcity is dominant in most countries, except for some countries. Most countries close to the equator (many in African regions and few countries in American region and West pacific region) appear to show high vulnerability to infectious disease in the context of freshwater consumption.

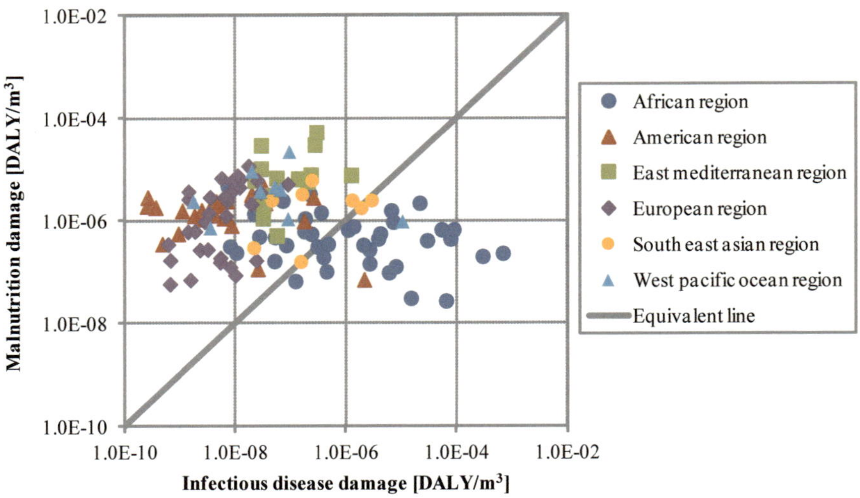

Figure 3. Comparison of infectious disease and malnutrition damage per unit volume freshwater consumption [13, 15]

All methods related to health damage assessment are not comparable because approaches and targets of the assessment are not perfectly corresponding with each other. However, methods of Pfister et al. [9] and Motoshita et al. [15] can be comparable in the aspect of malnutrition damage due to freshwater consumption. Malnutrition damage per freshwater consumption in both methods is plotted in Figure 4. Damage in the method of Motoshita et al. [15] seems to be larger than that of Pfister et al. [9] in most countries. The differences between both methods in the aspect of modelling procedures are selected parameters for reflecting social and economic condition and the consideration of ripple effects by international food trade. Same comparison is shown in Figure 5 after preliminarily excluding international food trade model in the method of Motoshita et al. [15]. Damage of both methods becomes much closer and the opposite tendency to Figure 4 can be seen in Figure 5. Thus, the effect of international food trade might be significant for the differences between both methods. The other method of Boulay et al. [8] cannot be simply compared with others because both of infectious and malnutrition damage are included and not separated as characterization factors. However, the scale of damage is not so different from that in other two methods.

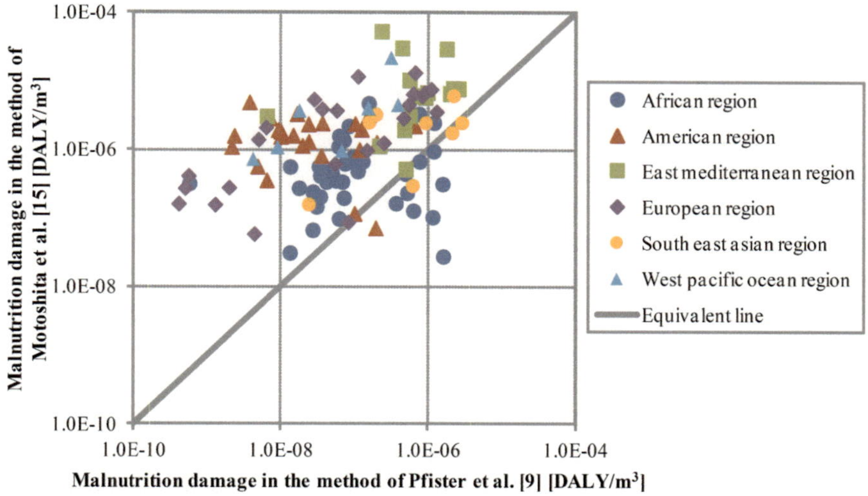

Figure 4. Comparison of malnutrition damage caused by freshwater consumption between in the methods of Pfister et al. [9] and Motoshita et al. [15]

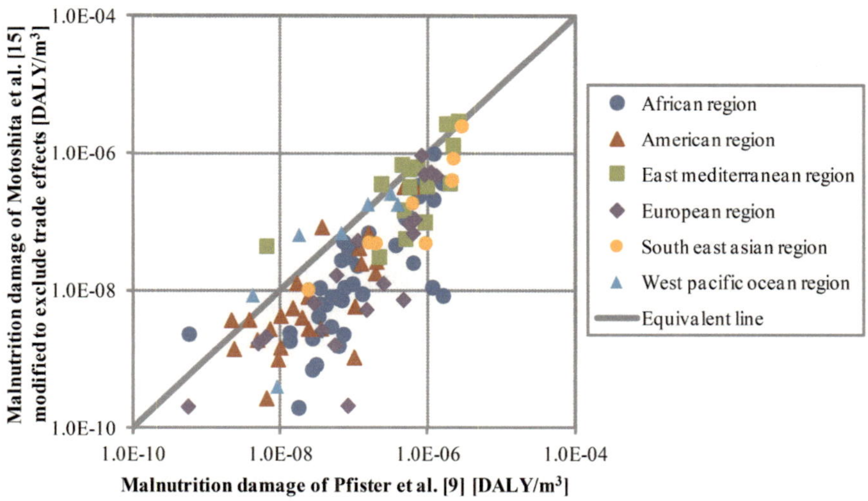

Figure 5. Comparison of malnutrition damage caused by freshwater consumption between in the methods of Pfister et al. [9] and modified Motoshita et al. [15] for excluding the effect of international food trade

3.2. Ecosystem

Freshwater resource is the essential not only for human but also ecosystem. Freshwater resource is utilized for sustaining life of living things and supplying habitats. Anthropogenic freshwater consumption may cause several types of effects on ecosystem. However, any consensus on cause-effect chain of freshwater consumption related to ecosystem has not been reached yet because of its complexity. On the other hand, several challenges on quantifying the part of impacts on ecosystem due to freshwater consumption have been made. Overview of them is introduced in the following sections.

Anthropogenic freshwater consumption will reduce the availability of freshwater for sustaining plant growth. Prevention of plant growth as a consequence of freshwater consumption was modelled by Pfister et al. [9]. In their modelling, the amount of net primary production (NPP) loss was calculated on grid scale for whole world by using the model calculating NPP limited by water availability [16]. Obtained NPP loss due to freshwater consumption was converted to vascular plant species biodiversity (VPBD) on the basis of the correlation analysis results between VPBD and NPP. Vascular plant species biodiversity was expressed by adopting the index of potentially disappeared fraction (PDF) used in Eco-indicator'99 [17]. While compensation by precipitation was considered in the model, the fate of freshwater from consumption to the availability loss for plants was very simplified by regarding that all the amount of consumed freshwater would restrict plant growth except for barren lands. Site specific water flow relevant to groundwater extraction was considered in the context of Netherland by van Zelm et al. [18]. The probability of occurrence of individual plant species was estimated by using the soil moisture indicator and the soil moisture could be described as a function of average groundwater level. The change of average groundwater level was modelled by hydrological zone model on grid scale. As a result, biodiversity loss of terrestrial plant species caused by groundwater extraction was quantified for the Netherland by using the indicator of potentially not occurring fraction of plant species (PNOF), which is almost same concept as PDF.

Consumption of freshwater may decrease habitats for aquatic species. Maendly et al. [19] modelled the effect of hydropower water dam on the number of aquatic species in downstream based on actual observed change of individuals of aquatic species due to dam construction. The effect of water demand for hydropower was express by adopting PDF. Generalized impact factor is proposed in the model, however it should be noted that the extrapolation was performed based on limited observation data (mainly in the context of Europe and United States of America).

3.3. Resources

Resources are determined as an endpoint of environmental load in life cycle impact assessment. However, "Resources" indicates very wide and fuzzy meanings. The safeguard subject relevant to "Resources" is dependent on methods due to their philosophy [17, 20, 21, 22]. In this context, damage on resources due to freshwater consumption has been quantified in different aspects.

For instance, depletion of fossil fuel or minerals will result in surplus energy demand for future generation to extract from lower grade resources [17, 20]. The same concept was adopted by Pfister et al. [9] for freshwater consumption. In the method, surplus energy demand for compensating the amount of consumed freshwater by desalination was evaluated as damage on resource only for the countries in that freshwater was overused compared with the available amount of freshwater. Surplus energy for compensation was calculated based on the state-of-the-art technology of desalination in the unit of MJ/m^3. Advantageous point of this method is high consistency with damage caused by consumption of other resources and fossil fuels [17, 20]. The significance of damage caused by resource consumption including freshwater consumption is comparable in the same unit (MJ).

Economic value of resources is also regarded as an endpoint of environmental impact and will be lost by resource consumption [21, 22]. In the same meaning, the economic loss of agricultural commodity due to agricultural water scarcity was quantified by Motoshita et al. [15]. The loss of agricultural commodity due to freshwater consumption was calculated based on crop productivity per unit volume of water on country scale and commodity price. In this context, animal commodity and aquacultural commodity should be also affected by freshwater consumption but did not considered in the method at present.

4. The specific example of the application to water footprinting

There are many kinds of methods from the perspectives of midpoint and endpoint as introduced in the above section. The specific example of the application will be helpful for understanding the significance of impact assessment in the context of water footprint. As an example, Pfister et al. [9] reported the results of impact assessment due to freshwater consumption in cotton textile production based on their method at midpoint and endpoint on country scale. The amount of freshwater consumption in 1kg cotton textile production and its impact at midpoint (shown as water deprivation) is shown in Figure 6. Generally, the impacts at midpoint level (physical scarcity) increase with the amount of consumed freshwater in Figure 6. However, some countries show relatively low impacts due to the physical abundance of available freshwater resources.

On the other hand, the impacts for each country at endpoint level are plotted against to those at midpoint level in Figure 7. The difference between physical stress of freshwater resources and specific results of water scarcity can be found out. For instance, Mali showed relatively lower impact than Australia in Figure 6, but human health damage as an impact at endpoint level is larger than Australia. While almost same amount of freshwater consumed for 1kg cotton textile production in both countries, the impacts at midpoint and endpoint shows opposite tendency. Thus, physical scarcity is not necessarily available for perfectly substituting for specific results of freshwater consumption.

The results of endpoint assessment on human health and ecosystem due to freshwater consumption for 1kg cotton textile production are shown in Figure 8. While human health damage due to freshwater consumption is relatively serious rather than damage on ecosystem in India and Mali, damage on ecosystem is more significant in Argentina, Australia and Mexico. The consequences of freshwater consumption are different among countries even though in the perspective of endpoint assessment.

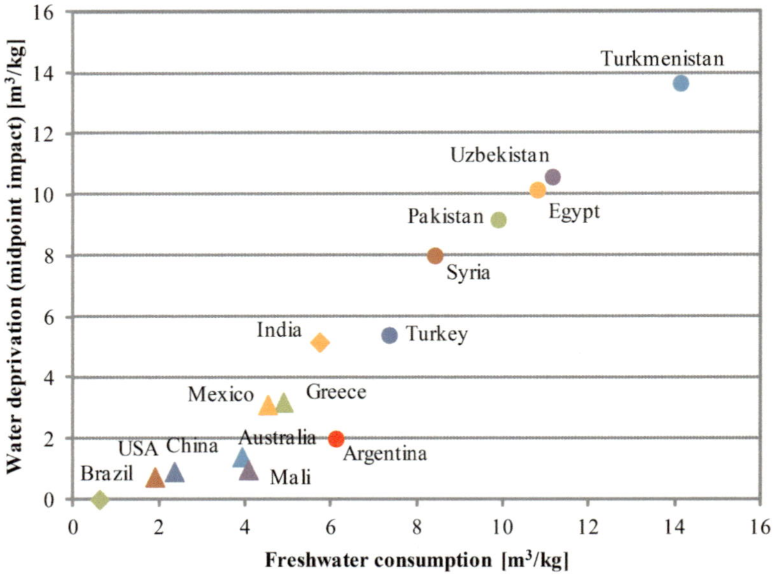

Figure 6. The comparison between the amount of freshwater consumption for cotton textile and its impact (water deprivation) at midpoint level

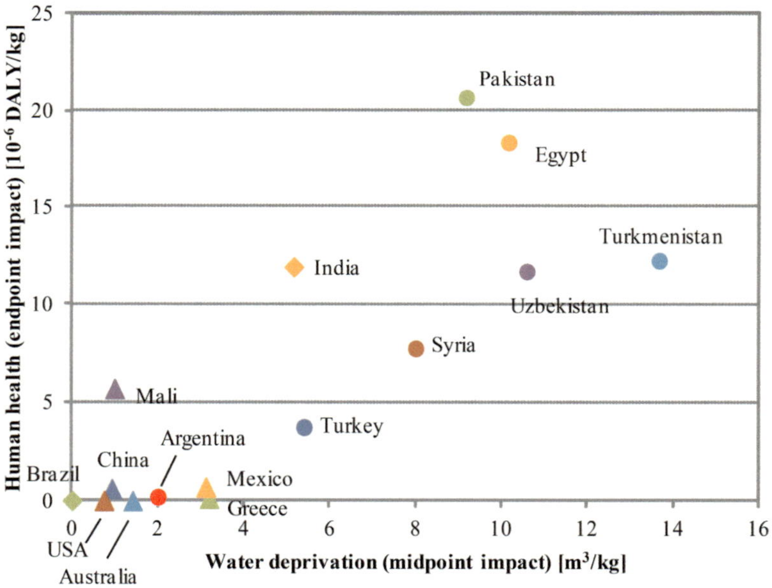

Figure 7. The comparison between water deprivation (midpoint impact) and human health damage (endpoint impact) due to freshwater consumption for cotton textile production

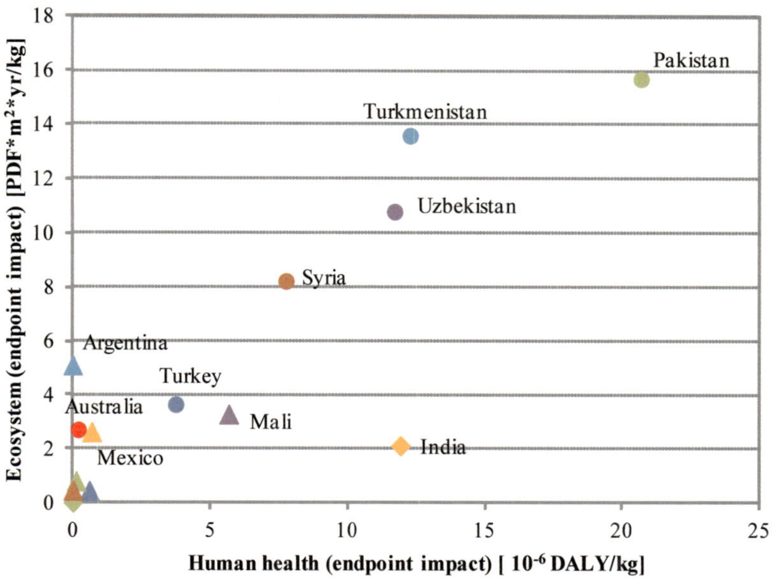

Figure 8. The comparison between damage on human health and ecosytem due to freshwater consumption for cotton textile production

5. Summary

As shown in the example of water footprinting, the amount of consumed freshwater is not an enough indicator to consider water stress in the quantitative aspect. There are many methods relevant to from midpoint to endpoint. Midpoint assessment is based on the physical scarcity and close to the cause side of freshwater consumption. The results of midpoint assessment have more robust relationship with freshwater consumption. On the other hand, endpoint models focus on the specific results of freshwater consumption and close to the effect side of freshwater consumption. Generally, uncertainty of the assessment results may increase in endpoint assessment due to considering the cause-effect chain of freshwater consumption. However, the assessment at endpoint level will make it possible to compare the effects of other environmental categories related to same endpoint. Therefore, water stress due to freshwater consumption should be assessed in both aspects of midpoint and endpoint. In addition, each assessment method has different characteristics on the basis of their philosophy. Sensitivity analysis by using multiple methods will be useful to verify the robustness of the assessment results. In recent years, many methods for quantifying water stress in the quantitative aspect have been developed. However, there is still more space to sophisticate the methods for more precise assessment and expand the targets of the modelling (ecosystem and resources in endpoint assessment).

Author details

Masaharu Motoshita*

Address all correspondence to: m-motoshita@aist.go.jp

Research Institute of Science for Safety and Sustainability, National Institute of Advanced Industrial Science and Technology, Tsukuba, Japan

References

[1] Shiklomanov I.A., Rodda J.C. - World Water Resources at the Beginning of the Twenty-First Century, Cambridge, Cambridge University Press; 2003.

[2] Chapagain A.K., Hoekstra A.Y. The water footprint of coffee and tea consumption in the Netherlands, Ecological economics 2007; 64 109-118.

[3] Humbert S., Loerincik Y., Rossi V., Margni M., Jolliet O. Life cycle assessment of spray dried soluble coffee and comparison with alternatives (drip filter and capsule espresso), Journal of Cleaner Production 2009; 17 1351-1358.

[4] Owens J. W. Water resources in life-cycle impact assessment: Considerations in choosing category indicators, Journal of Industrial Ecology 2002; 5(2) 37-54

[5] Frischknecht R., Steiner R., Arthur B., Norbert E., Gabi H. Swiss Ecological Scarcity Method: The New Version 2006, ESU-Services; 2006. http://www.esu-services.ch/fileadmin/download/Frischknecht-2006-EcologicalScarcity-Paper.pdf (accessed 14 April 2011)

[6] Mila i Canals L., Chenoweth J., Chapagain A., Orr S., Anton A., Clift R. Assessing freshwater use impacts in LCA: Part. I — inventory modelling and characterisation factors for the main impact pathways, International Journal of Life Cycle Assessment 2009; 14 28-42

[7] Hoekstra, A.Y. and Mekonnen, M.M. Global water scarcity: monthly blue water footprint compared to blue water availability for the world's major river basins, Value of Water Research Report Series No. 53, Delft: UNESCO-IHE; 2011. http://www.waterfootprint.org/Reports/Report53-GlobalBlueWaterScarcity.pdf (accessed 19 August 2012)

[8] Boulay A. M., Bulle C., Bayart J. B., Deschenes L., Margni M. Regional characterization of freshwater use in LCA: modelling direct impacts on human health, Environmental Science and Technology 2011; 45 8948-8957

[9] Pfister S., Koehler A., Hellweg S., Assessing the Environmental Impacts of Freshwater Consumption in LCA, Environmental Science and Technology 2009; 43 4098-4104

[10] Ridoutt B. G., Pfister S., A revised approach to water footprinting to make transparent the impacts of consumption and production on global freshwater scarcity, Global Environmental Change 2010; 20(1) 113-120

[11] Food and Agriculture Organization of the United Nations, FAO: AQUASTAT. http://www.fao.org/nr/water/aquastat/main/index.stm (accessed 19 August 2012).

[12] Prüss-Üstün A., Bos R., Gore F., Bartram J., Safer water, better health – costs, benefits and sustainability of interventions to protect and promote health. Geneva: World Health Organization; 2008. http://whqlibdoc.who.int/publications/2008/9789241596435_eng.pdf (accessed 19 August 2012).

[13] Motoshita M., Itsubo N., Inaba A., Development of impact factors on damage to health by infectious diseases caused by domestic water scarcity, International Journal of Life Cycle Assessment 2010; 16(1) 65-73

[14] Watkins K., Human Development Report 2006. New York: United Nations Development Programme; 2006. http://hdr.undp.org/hdr2006/pdfs/report/HDR06-complete.pdf (accessed 19 August 2012).

[15] Motoshita M., Itsubo N., Inaba A., Damage assessment of water scarcity for agricultural use: proceedings of 9th conference on EcoBalance, 9-12 November 2010, Tokyo, Japan. Tokyo: The Institute of Life Cycle Assessment, Japan; 2010.

[16] Nemai R. R., Keeling C. D., Hashimoto H., Jolly W. M., Piper S. C., Tucker C. J., Myneni R. B., Running S. W., Climate-driven increases in global terrestrial net primary production from 1982 to 1999, Science 2003; 300(5625) 1560-1563

[17] Goedkoop M., Spriensma R., The Eco-indicator 99 A damage oriented method for Life Cycle Impact Assessment: Methodology Annex. Amersfoort: PRe consultans; 2001. http://www.pre-sustainability.com/download/misc/EI99_annexe_v3.pdf (accessed 19 August 2012)

[18] van Zelm R., Schipper A. M., Rombouts M., Snepvangers J., Huijbregts M. A. J., Implementing groundwater extraction in life cycle impact assessment: Characterization factors based on plant species richness for the Netherlands, Environmental Science and Technology 2011; 45 629-635

[19] Maendly R., Humbert S., Empirical characterization model and factors assessing aquatic biodiversity damages of hydropower water use, International Journal of Life Cycle Assessment, submitted.

[20] Goedkoop M., Heijungs R., Huijbregts M., De Schryver A., Struijs J., van Zelm R., ReCiPe 2008: A life cycle impact assessment method which comprises harmonised category indicators at the midpoint and the endpoint level. : ReCiPe; 2012. https://sites.google.com/site/lciarecipe/file-cabinet/ReCiPe_main_report_REVISED_13-07-2012.pdf?attredirects=0 (accessed 19 August 2012)

[21] Itsubo N., Inaba A., LIME2: Life-cycle Impact assessment Method based on Endpoint modeling. Tokyo: Life-Cycle Assessment Society of Japan; 2012. http://lca-forum.org/english/pdf/No14_C1_Outline.pdf (accessed 19 August 2012)

[22] Steen B., A systematic approach to environmental strategies in product development (EPS), Version 2000 – Models and data of the default methods. Gothenburg : Chalmers University of Technology; 1999.

Chapter 2

Drought and Its Consequences to Plants – From Individual to Ecosystem

Elizamar Ciríaco da Silva,
Manoel Bandeira de Albuquerque,
André Dias de Azevedo Neto and
Carlos Dias da Silva Junior

Additional information is available at the end of the chapter

http://dx.doi.org/10.5772/53833

1. Introduction

Climate-change scenarios around the world indicate that many areas of the globe will increase in aridity. Thus, all living organisms will suffer from a water scarcity, especially plants, which do not have locomotive structures that allow them to move elsewhere when water and food becomes scarce. As a result, different terrestrial ecosystems (natural and agricultural) will be severely affected and some may even collapse due to the extinction of plant species.

It is therefore important to gain a better understanding regarding the effect of frequent drought stress on biochemical and physiological processes in plants as well as on the plant population and/or community in a particular ecosystem. Despite the negative aspects of such changes, severe environmental conditions can induce interesting adaptations in plants that allow them to survive and reproduce. These adaptations can lead to the emergence of new functional groups in a given ecosystem or serve as an important tool for improving agricultural practices and plant breeding programs.

In recent decades, a large number of investigations have addressed strategies used by plants to control water status, avoid oxidative stress and maintain vital functions in an attempt to understand the morphological and physiological changes plants undergo to ensure their survival under different environmental conditions. Special attention has been given to molecular processes involved in drought tolerance and resistance. While some advances have been made, we still do not fully understand the underlying survival mechanisms in plants due the complex interaction of different forms of stress in natural habitats.

On the ecosystem level, drought induces changes in different processes and frequently demands functional plant responses. Some ecosystems, such as savannas, steppes and scrublands, have intermittent low annual precipitation. In these water-limited ecosystems, drought can seasonally modify carbon and nitrogen cycles, resulting in poor water and mineral uptake by roots, lesser plant growth, a reduction in litter decomposition and the biogenic emission of CO_2 from the soil. Severe drought can also induce a higher vegetation mortality rate due to cavitation and/or carbon starvation (reduced photosynthesis and enhanced autotrophic respiration). Thus, more frequent and intense drought periods (and the consequent death of plant species) can alter the phytosociology of entire plant communities over time.

Reductions in aboveground net primary productivity and alterations in functional plant groups are observed in places subjected to prolonged, severe drought. This chapter offers an overview of the effect of drought on individual plants and ecosystems, emphasising aspects of growth, water relations and photosynthesis, especially the electron transport chain, as well as radical oxygen species (ROS) scavenging and its role in avoiding oxidative stress. On the ecosystem level, functional traits commonly associated to water stress tolerance and changes in ecological processes and functional responses in plants will be also discussed.

2. Drought as a stress factor to the plants

In recent decades, a large number of models have been developed to estimate climate changes around the world. Climate change is defined as a significant difference between two mean climatic states, with substantial impact on the ecosystem [1]. Extreme climatic events, such higher temperatures, more intense precipitation, increased drought risk and duration as well as cyclones and flooding in coastal areas, are expected to increase in both frequency and intensity [2, 3]. In some countries, large arid and semiarid areas are expected to increase in size, leading to desertification. Currently, the consequences of global warming are widely discussed, especially regarding plant productivity and the increase in areas subject to desertification.

According to Assad et al. [4], the average temperature of the planet will increase by 1.4 to 5.8 °C by the end of the century, with drought being one of the consequences of this warming. Thus, one may deduce that the planet is heading toward a serious water crisis. Desertification corresponds to a reduction in the productive capacity of arid, semiarid and sub-humid lands as a result of climatic and edaphic factors. This growing, worldwide phenomenon has been causing both social and environmental problems, including the disappearance of animal and plant species [5].

In semiarid regions of Brazil, for example, inappropriate cultivation techniques resulting in soil erosion and a loss of water retention capacity in the soil as well as the expansion of livestock farming and the indiscriminate extraction of firewood deplete the nutritional content of the soil, thereby contributing toward the process of desertification. These activities lead to progressive degradation that results in the loss of soil cover [6, 7].

Plants need a large amount of water and nutrients throughout their life cycle and all aspects of plant development are affected by a reduction in water content in the soil. This reduction in soil moisture leads to changes in the physical environment, which subsequently affect physiological and biochemical processes in plants [8-10]. Drought can cause nutrient deficiencies, even in fertilised soils, due the reduced mobility and absorbance of individual nutrients, leading to a lower rate of mineral diffusion from the soil matrix to the roots [3]. Thus, drought is doubtlessly the most important stress factor limiting plant life.

Water is required for processes such as germination, cell division and elongation for the promotion of plant growth in height and width and metabolic activities, such as the synthesis of organic compounds, photosynthesis, respiration and a number of other physiological and biochemical processes [11]. Thus, when water availability decreases, changes occur in all molecular, biochemical, physiological and morphological aspects of plants.

Drought triggers a wide variety of plant responses [12]. Plant growth is altered, with changes in the architecture of individuals, which are translated into lower height, reduced leaf size, a smaller number of leaves, less fruit production and changes in the reproductive phase. Osmoregulatory processes generally occurs to protect membrane integrity and maintain the inflow of water to the cell as well as the accumulation of organic solutes as sugars, quaternary ammonium compounds (glycine betaine and alanine betaine) [13, 14], hydrophilic proteins (late embryogenesis abundant proteins) [15], soluble proteins and amino acids (proline) [10, 14]. Water is the most important substance in the initial phase of plant development from germination and seedling formation to establishment in the field [16] and a reduction in the water supply in this stage can lead to dehydration and even death.

In agricultural ecosystems, drought has a detrimental effect on crop production, affecting the growth rate and development of the economically important portions of the plant, such as fruits, grains and leaves. Without irrigation, production in crops such as coffee can be reduced by as much as 80% in dry years [17]. In Mexico, 80% of the problems caused by drought are related to losses in agricultural systems [18]. During a 45-day drought in the state of Paraná, Brazil, the 2008/2009 soybean harvest was reduced by 80% in areas without dry cover [19]. The same can be estimated for important crops such as sugarcane, corn, wheat and a number of others. The tragic effect on productivity is explained by the vital importance of water in living cells, which affects all biochemical and metabolic processes.

2.1. Water relations and influence on plant growth and development

Water is attracted to soil pores predominantly due to its attraction to other surfaces (adhesion) and capillarity. Its movement in the soil occurs mainly through mass flow: water fills micropores in the soil, which are interconnected and allow water movement. Contact between the surface of the roots (mainly in the root hair zone) and soil provide the surface area necessary for water uptake. The growth of the roots into the soil maximises water absorption [11].

Water flow from the soil to the roots depends on the water potential gradient between the soil and plant, which is affected by the water needs of the plant, the hydraulic conductivity of the soil, soil type, moisture content in the soil [20] and the atmospheric demand, which directly affects water loss through transpiration, generating considerable tension in the xylem and contributing to the creation of this potential gradient. Water potential (Ψw) is an expression of the energy status of water in any system, such as soil, tissues, the whole plant or the atmosphere, and its energy is influenced by four components: surface force or matrix potential (Ψm), gravitational potential (Ψg), hydrostatic pressure or pressure potential (Ψp) and solutes or osmotic potential (Ψs), which, in most cases, exert a negative effect on total water potential (Ψw), reducing water energy and consequently water capacity for moving into a system. Thus, water flow in the soil-plant-atmosphere system always follows a downhill direction from higher to lower, which is the driving force of water transport [11, 20]. Water potential is always represented by negative values. The reference is pure water under normal conditions of temperature and pressure assumed to be equal to zero ($\Psi w = 0$), which denotes maximum energy status.

In wet soil, the hydrostatic pressure is closer to zero and Ψw is about -0.03 MPa [11]. A reduction in the water supply when the soil becomes dry leads to a decrease in hydrostatic pressure (Ψp), which becomes quite negative. Thus, due to the high surface tension that tends to minimise the air-water interface, water becomes strongly adsorbed by the electrical charges of the soil particles (adhesion) [11, 20]. Under this condition, the plant absorption process requires a reduction in Ψw in the roots cells in relation to the rhizosphere. Moreover, the constant absorption of water by the plant leads to a reduction in the moisture content of the neighbouring soil.

The coordination of water flow from the soil to the roots, xylem, leaf apoplast and bulk air follows a decreasing status of water energy. This water gradient established between the rhizosphere through the plant and atmosphere favours the inflow of water in well-watered plants. In dry soil, however, the flow is interrupted due to barriers in the soil, such as increased surface forces, as well as in the plant, such as resistance offered by stomatal closure [20, 21]. When moisture availability in the soil decreases and there is continuity in the loss of water through transpiration, cavitation can occur, causing the interruption of water flow through the xylem due to the formation of air bubbles.

The continued inflow of water contributes to growth processes, as turgor pressure is responsible for cell elongation. Plant growth is the result of daughter-cell production by meristematic cell divisions in the shoot and root and the subsequent massive expansion of the young cells [12]. The constant inflow of water exerts pressure within the cell, causing the cell wall to stretch and inducing the elongation or growth of the cell in both size and volume. This physical process is repeated until the cell becomes mature, at which point cell size is no longer significantly altered [11]. These two processes (cell division and expansion) are important to the growth and development of tissues and organs.

Dry soil and the loss of water through a high transpiration rate makes the plant experience drought stress [12], which leads to the loss of turgor. As a result, the development of some structures is compromised and the growth rate slows. Thus, plants are generally shorter in dry environments. Although the formation of the organs is genetically defined, environmental conditions exert an influence on this process. Once formed, the cells of the leaves and fruit rarely undergo cell division and their growth relies on cell expansion. If the water pressure is insufficient to promote elongation, these organs will be small in relation with the those formed in a well-hydrated environment [22].

Plants also need carbon dioxide and light to produce organic matter throughout the process of photosynthesis. Carbon dioxide enters the leaves through the stomata and the turgor of the guard cells is the regulatory mechanism for maintaining the stomata opened [11]. Plants differ morphologically and/or physiologically under drought conditions. Different mechanisms allow plants to survive and even produce with a limited water supply, such as the maximisation of water uptake by deep, dense root systems, the minimisation of water loss by stomatal closure and a reduction in leaf area, osmotic adjustment or changes in cell wall elasticity as well as other essential processes for maintaining physiological activities throughout extended periods of drought [23].

Deciduous species have an efficient mechanism for coping with drought, which involves stomatal closure, changes in the orientation of the leaf and the reduction in leaf area by shedding leaves to minimise water loss through transpiration [24]. In the dry season, the leaves that remain on the plant can strongly influence the water balance by adjusting transpiration as a function of hydraulic limitation due to an increase in atmospheric vapor pressure deficit and surface soil desiccation [25].

Cell turgor is maintained by the accumulation of organic substances and inorganic ions in a stress response mechanism denominated osmotic adjustment [26, 27]. Organic solutes, also referred to as compatible solutes, are highly soluble compounds of low molecular weight that have no toxicity at high concentrations within the cells [14]. When plants are exposed to water deficit, changes occur in biochemical substances, such as the conversion of starch to soluble sugars (sucrose, glucose, fructose, etc.) [9, 27,28]. Nitrogenous compounds, such as proteins, amino acids (arginine, proline, lysine, histidine, glycine, etc.) and polyamines, are another group of compounds affected by water deficit that participate in osmotic adjustment [29]. In response to drought, there is an increase in the levels of free amino acids [9] and a reduction in the rate of synthesis or a decrease in proteins [29]. The increase in proline content is of considerable importance to plant adaptation during stress [8] and its accumulation usually occurs in large amounts in higher plants in response to environmental stress [14]. Proline is an amino acid resulting from the hydrolysis of proteins and plays an important role as an osmoprotectant in many cultivated species [27, 28, 30]. The increase of proline has also been linked to the reduction in leaf water potential [30]. In addition to its role as an osmoregulator, proline stabilises membranes and proteins and contributes to the removal of free radicals [14].

3. Drought and photosynthesis

Drought is arguably the most important factor limiting plant yields throughout the world. Climate change and global warming in the tropical zone is expected to affect the photosynthesis, development and biomass production of plants in many regions as a result of the significant rise in temperature and concentration of atmospheric CO_2, which will also lead to a reduction in water availability in the soil, with a consequent effect on carbon assimilation and plant growth [31]. Semiarid regions are subject to water shortages and soil degradation in such places is likely to increase with climate change. The response of photosynthesis to drought merits special attention, as water is an electron donor that allows the maintenance of this process and biomass productivity [32, 33].

Under conditions of low water availability, a reduction in stomatal conductance constitutes one of the first strategies used by plants to diminish the transpiration rate and maintain turgescence [34]. Accordingly, stomatal behaviour in response to situations of drought stress may be indicative of water use efficiency for the production of photosynthates. Exposure to stress may induce alterations in photobiological processes, resulting in stomatal restrictions regarding the supply of carbon dioxide, the loss of water vapour and limitations to non-stomatal components, with harm to the reaction centres of photosystems I and II (PSI and PSII), thereby compromising photosynthesis efficiency [32]. According to Bolhàr-Nordenkampf et al. [35], Bolhàr-Nordenkampf and Öquist [36] and Baker [37], changes in the photochemical efficiency of plants under drought conditions may be assessed through an analysis of chlorophyll *a* fluorescence efficiency associated with PSII.

The chlorophyll fluorescence of water-stressed barley plants is characterised by a mild decrease in Fv/Fm (Fv is the variable part of Chl fluorescence and Fm is Chl fluorescence intensity at the peak of the continuous fluorescence inductive curve) and significant increase in F0 (Chl fluorescence with all PSII reaction centres open), together with a slight decrease in Fm [38]. The optimal temperature for most species ranges from 25 to 35 °C, above which a decline in the rate of photosynthesis is observed [39, 40]. Under natural conditions, momentary water deficit is observed during warm hours of the day, which promotes stomatal closure. Consequently, the temperature of leaves exposed to direct sunlight can be equal to or higher than the air temperature. This rise in leaf temperature results in biochemical and biophysical disturbances in the mesophyll, which may or may not be reversible [39].

The main effects of high temperature on photosynthesis result from alterations in thylakoid physical-chemical properties [41], besides inducing an increase in lipid matrix fluidity [42], with the consequent formation of a single-layer structure. High temperature causes the following disturbances to the organisation of the photosynthetic apparatus: a) destruction of the oxygen evolution complex; b) dissociation of the light harvesting complex of PSII accompanied by variations in energy distribution between PSII and PSI; and c) inactivation of the PSII reaction centre (P680), which disturbs grana stacking [43]. All these events result in the loss of photochemical and carboxylation efficiency as well as serious metabolic restrictions in the Calvin cycle, such as the inactivation of ribulose-1,5-bisphosphate carboxylase/ oxygenase and variations in the metabolic pool, especially ATP and NADPH availability

[44]. In some situations, F0 can be used as an indicator of irreversible damage to PSII [45] associated with LHCII dissociation [43, 46] and the blocking of the electron transference on the reductant side of PSII. In wheat and barley plants, high temperature tolerance is positively correlated with maximum F0 [47]. However, Yamane et al. [48] suggest that the inactivation of the PSII reaction centre caused by the denaturation of chlorophyll-protein complexes in response to high temperature correlates with a decay in Fm values. Changes in these fluorescence variables cause alterations in the Fv/Fm ratio, indicating a disturbance in the photochemical activity of photosynthesis. The Fv/Fm ratio has been inferred as an indicator of environmental stress, such as high temperature, drought and excess light, as it is easy and fast to measure [49].

3.1. Aspects of chlorophyll *a* florescence transient: *Kielmeyera rugosa* Choisy as case study

The genus Kielmeyera belongs to the family Clusiaceae (Guttiferae), subfamily Kielmeyeroideae, and is endemic to South America. The vast majority of these species occur exclusively in Brazil, where nearly 50 species are found chiefly in the *restinga* (sand dune), rocky savannah and the savannah-like *cerrado* vegetation south of the Amazon [50]. Some species are traditionally used in Brazilian folk medicine to treat tropical diseases, such as schistosomiasis, leishmaniasis and malaria, as well as fungal and bacterial infections [51].

A case study was performed with a population of 10 adult plants of *Kielmeyera rugosa* Choisy (Clusiaceae) in a *restinga* ecosystem in the municipality of Pirambu, state of Sergipe (northeastern Brazil), where the climate is characterised by irregular rainfall, with a wet season from April to September. Leaf water potential (Ψw) was determined between 9:00 and 11:00 am and the chlorophyll and chlorophyll *a* fluorescence indexes were determined between 12:00 and 1:00 pm in March 2011 (dry season) and July 2011 (wet season). The mean air temperature in the rainy and dry seasons was 26.8 and 39 ºC, respectively.

Chlorophyll transient florescence (JIP-test): Polyphasic Chl *a* florescence transient (OJIP) was measured in healthy, completely expanded leaves using a hand-held fluorometer (Handy-PEA, Hansatech, King Lynn, UK). The selected leaves were subjected to a 30-min dark adaptation period, which is enough time for all reaction PSII centres to open [52]. The leaves were then immediately exposed to a pulse of saturating light at an intensity of 3000 μmol.m-2s^{-1} provided by an array of three high-intensity light-emitting diodes. The JIP-test [53] was used to analyse each Chl *a* fluorescence transient. This test is based on the energy flux from bio-membranes [54]. The performance index (PIABS) [55] was employed as a parameter to quantify the effects of environmental factors on photosynthesis in several studies.

Figure (1A) shows that *K. rugosa* underwent a significant decrease of 120 and 38% in leaf water potential and the chlorophyll index (1B), respectively, in the dry season. Mean leaf Ψw was -0.34 MPa in the wet season and -0.75 MPa in dry season.

An analysis of florescence transients in *K. rugosa* under the two distinct water availability conditions (wet and dry season) may provide information on changes taking place in the structure, conformation and function of the photosynthetic apparatus, especially in PSII. Initial florescence (F0) represents the basal emission of Chl florescence when redox components of photosystems are fully oxidised. This requires appropriate dark adaptation. The results reveal an increase in F0 in the dry season, which may be explained by the initial damage occurring in PSII, likely due to the high temperatures and low water availability (Table 1). This increase in F0 is dependent on structural conditions affecting the probability of the energy transference within the pigments of the light harvesting complex to the PSII reaction centre [56]. According to Bolhàr-Noderkampf et al. [35], the increase in F0 increase in the dry season may indicate a reduction in energy transference to the PSII reaction centre or a partially-reversible inactivation [48].

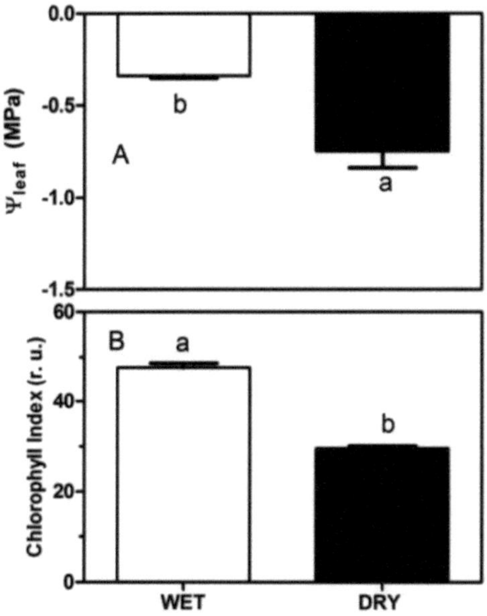

Figure 1. Mean values of leaf water potential (A) and Chlorophyll index (B) on wet and dry season in *Kielmeyera rugosa* Choisy growing under field conditions at 'restinga' in the Municipality of Pirambu, Sergipe State, Brazil. Each value represents a means of 10 replicates and bars indicate standard deviations. Mean values followed by the same small letters for the seasons are not significantly different (*P>0.05*; t-test). (Silva Junior CD, unpublished data).

The strong decrease in Fm in the dry season was likely associated with the higher temperatures (Table 1). This decrease in *K. rugosa* may be related to the loss of PSII activity due to conformational changes in the D1 protein [57], causing alterations in the properties of PSII electron acceptors [58]. Other factors may be associated with the heat-related decrease in Fm, such as the migration of damaged PSII reaction centres to non-stacked thylacoid regions and accelerated energy transference to non-fluorescent PSI [48]. The decrease in Fm may also be due to the disruption of electron donation from water to PSII due to the loss of the manganese atom and extrinsic proteins from the oxygen evolution complex [59]. Such events may be related to susceptibility to high temperatures.

	Seasons	
Variable	**Wet**	**Dry**
F0	513 ± 7b	627 ± 26a
F50μs (O)	570.5 ± 9b	692.8 ± 27a
100 μs	622.3 ± 12b	760.5 ± 30a
F300μs	840.4 ± 19b	966.2 ± 41a
F2ms (J)	1434 ± 32a	1299 ± 53b
F30ms (I)	2394 ± 67a	1575 ± 88b
Fv	2425 ± 59a	1352 ± 123b
Fm (P)	2938 ± 65a	1979 ± 110b
tFm	370.0 ± 26a	248,0 ± 12b
Area	**67636 ± 2308a**	**35236 ± 2657b**

Table 1. Initial florescence (F0), florescence intensity at 50 μs ($O=F_{50\mu s}$), 100 μs ($F_{100\ \mu s}$), 300 μs ($F_{300\mu s}$), 2 ms ("J"$_{F2ms}$), and 30ms ("I"$_{F30ms}$), variable florescence (F_v) maximum florescence (F_m=P) time to reach Fm (t_{Fm}) and area beneath the florescence in *Kielmeyera rugosa* Choisy on wet and dry season. Mean values (n=10) ±SE are show. Mean values followed by the same small letters for the seasons are not significantly different (*P>0.05*; t test). (Silva Junior CD, unpublished data).

The area over the fluorescence curve between F0 and Fm was lower in dry season than in the wet season, suggesting a decrease in the electron pool size of PSII, including QA, QB and PQ (Table 1) [60]. If the electron transfer from the reaction centre (RC) to the quinone pool is blocked, this area is dramatically reduced [61]. In comparison to the wet season, the area over the florescence curve was significantly decreased with the increase in drought and temperature. This inhibition is more accentuated by the interaction between high temperatures and light intensity, which leads to the blockage of electron transfers from the RC to the quinone pool. These results are in agreement with those described by Metha et al. [62], who found an inhibition in the electron transfer rates on the donor side of PSII in *Triticum aestivum* leaves treated with 0.5 M NaCl.

The results of flux ratio (yields) in *K. rugosa* revealed a decrease in TRO/ABS (φ_{Po}), ETO/TRO (ψ_o) and, consequently, ETO/ABS (φ_{Eo}) in the dry season (Figure 2 A, E and B). The decrease in φ_{Po} (18%) under water stress indicates a loss of the maximum quantum efficiency of primary photochemistry due to photoinhibition caused by excess energy. Moreover, this excess induced the inactivation of 31% of active RCs per cross-section in the dry season, causing increased energy dissipation as well as lower φ_{Po} values (Figure 2C). Under water stress, *K. rugosa* also exhibited a 35% decrease in φ_{Eo} in comparison to the wet season.

The performance index (PIABS) combines three independent functional steps of photosynthesis (the density of RCs in the chlorophyll bed, excitation energy trapping and conversion of excitation energy to electron transport) in a single multi-parametric expression [55], which is a function of ψ_0, φ_{Po} and RC/ABS [63, 64]. The results revealed much higher PIABS values in the wet season than in the dry season, possibly due to the photoinhibition caused by excess of light energy and lower water potential (Figure 1).

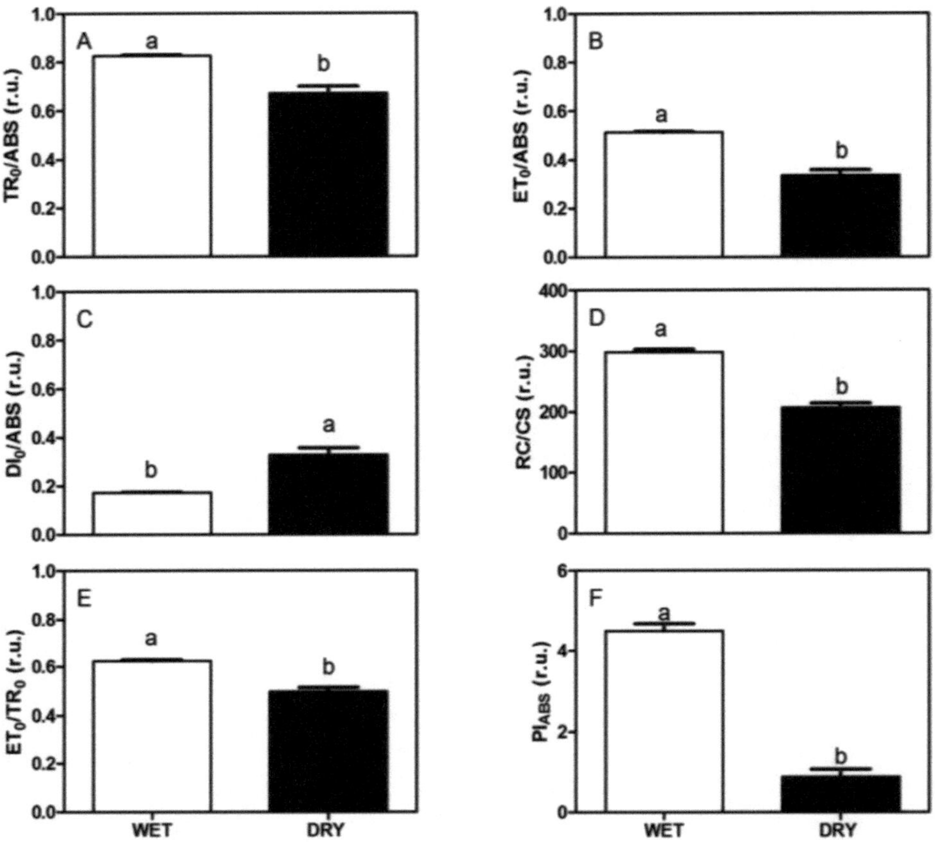

Figure 2. Maximum efficiency of PSII (φ_{Po}= TR_0/ABS), maximum efficiency of non-photochemical de-excitation (φ_{Do}= DI_0/ABS), probability that a trapped exciton (ψ_o= ET_0/TR_0) or that an absorbed photon (φ_{Eo}= ET_0/ABS) can move an electron further from QA, density of active reaction centers per cross section (RC/CS), and performance index (PIABS) in *Kielmeyera rugosa* Choisy under wet and dry season. Mean values followed by the same small letters for the seasons are not significantly different (*P>0.05*; t test). Mean values (*n*=10) ±SE. (Silva Junior CD, unpublished data).

φ_{Po} (Fv/Fm = TRO/ABS) is a parameter that expresses maximal PSII efficiency, which is controlled by the primary photochemistry of PSII (charge separation, recombination and stabilisation), the non-radiative loss of excited states in light-harvesting antennae and excited states quenched by oxidised PQ molecules from the PQ pool [65]. The low φ_{Po} values in *K. rugosa* under drought conditions could have resulted from the inactivity of the RCs, which may favour greater energy dissipation in the form of heat and fluorescence, as deduced from the high φ_{Do}values. This may be associated with increased heat sinks (heat-sink centres or silent centres), which may absorb light in a similar manner as that of active RCs, but are unable to store the excitation energy as redox energy and dissipate their total energy as heat

[66]. Moreover, due to excess irradiance, the transfer of energy to other systems could also take place, such as the energy-dependent formation of ROS [61].

Analysing Ψ_0, the lowest φ_{Po} values in *K. rugosa* were found under drought conditions. Ψ_0 values decreased to a remarkably greater extent in the dry season in comparison to wet season. This result reflects a reduction in the pool of plastoquinone (PQ) in an oxidised state and the reoxidation inhibition of QA- and demonstrates that, besides the loss of energy to QA, the loss of excitation energy further from QA was significant [67]. The φ_{Po}, Ψ_0 and φ_{Eo} results allow one to deduce that *K. rugosa* may use light energy more efficiently in the wet season due to the greater amount of chlorophyll and higher leaf water potential (Figure 1A,B).

The performance index (PIABS) is a consistent parameter for the evaluation of plant performance regarding light energy absorption, excitation energy trapping and the conversion of excitation energy to electron transport by photosynthesis under different stress conditions [55, 68]. The PIABS expresses both a function of the fluorescence extreme F0 and Fm as well as the intermediate J-step and the slope at the origin of the fluorescence rise, whereas φ_{Po} expresses a function of only F_0 and Fm, independently of how the trajectory of the fluorescence intensity reaches its maximal value [68]. Furthermore, the PIABS allows a broader analysis of photosynthetic performance, such as the relationship between photon absorption efficiency and the capture of excited energy in PSII, as well as an analysis of the density of active RCs and the probability that excited energy moves an electron further than QA-. Therefore, the PIABS is a better parameter for evaluating the responses of PSII to stressful conditions than φ_{Po} alone. In the present case study, the PIABS in *K. rugosa* was much lower in the dry season.

4. Oxidative stress and its effect to plants

4.1. Living with oxygen

The production of reactive oxygen species (ROS) is an unavoidable consequence of life with oxygen. The introduction of molecular oxygen (O_2) in the atmosphere during the Paleoproterozoic era (between 2.7 billion and 1.6 billion years ago) by the emergence of photosynthetic blue-green algae and later by higher plants led to the accumulation of O_2 in the atmosphere and oceans, inducing substantial changes in the living conditions of the earth. The atmosphere gradually changed from a reducing to an oxidising environment, thereby altering the pace and direction of evolution [69]. Ever since, ROS have been the unwelcome companions of aerobic life. Unlike of O_2, these partially reduced or activated derivatives of oxygen [singlet oxygen ($^1O_2$), superoxide radical ($^{\bullet}O2^-$), hydrogen peroxide (H_2O_2) and hydroxyl radical ($^{\bullet}OH$)] are highly reactive and toxic and can cause oxidative damage to carbohydrates, lipids, amino acids, proteins and nucleic acids [70]. Consequently, the evolution of all aerobic organisms has been dependent on the development of efficient ROS-scavenging mechanisms.

Under normal plant growth conditions, ROS are continuously produced and scavenged in organelles, such as chloroplasts, mitochondria and peroxisomes. However, the balance between ROS-producing pathways and ROS-scavenging mechanisms may be disrupted when plants experience environmental stress, such as drought, flooding, salt, heat, chill, heavy metals, nutrient deficiencies, UV radiation, intense light, air pollutants, herbicides, mechanical stress and attacks from pathogens [71].

The excessive production of ROS is responsible for secondary stress known as oxidative stress. Therefore, plant tolerance to drought and other forms of abiotic stress that induce an increase in the generation of ROS depends on the development of efficient ROS-scavenging mechanisms.

4.2. Chemistry of ROS

Much of the behaviour of molecular oxygen (or dioxygen) and its partially reduced species derive from their reduction potentials and molecular orbital structures. The dioxygen molecule is a highly unusual, stable diradical with a pair of electrons with parallel spins. To oxidise a non-radical atom or molecule, dioxygen would need to react with a chemical species that provides a pair of electrons with parallel spins that fit into its free electron orbitals. Fortunately, pairs of electrons typically have opposite spins, which imposes a restriction on the reaction of molecular oxygen with most organic molecules, such as amino acids and nucleic acids [70].

However, dioxygen may be converted into ROS either by energy transfer or monovalent reduction. If oxygen absorbs enough energy to reverse the spin of one of its unpaired electrons, it forms singlet oxygen ($^1O_2$), in which the two electrons have opposite spins. Since paired electrons are common in organic molecules, singlet oxygen is much more reactive toward organic molecules than dioxygen in its ground state. The second mechanism of oxygen activation is stepwise monovalent reduction through electron transfer reactions with the unpaired electrons of transition metals and organic radicals, resulting in the sequential formation of superoxide anion ($^{\bullet}O2^{-}$), hydrogen peroxide (H_2O_2), hydroxyl radical ($^{\bullet}OH$) and, finally, water (Figure 3). The first reduction step is free energy dependent (endergonic) and requires electron donation, but the following one-electron reduction steps are exergonic and can occur spontaneously, using transition metal ions (Fe^{2+} and Cu^{+}) and semiquinones as electron donors [70].

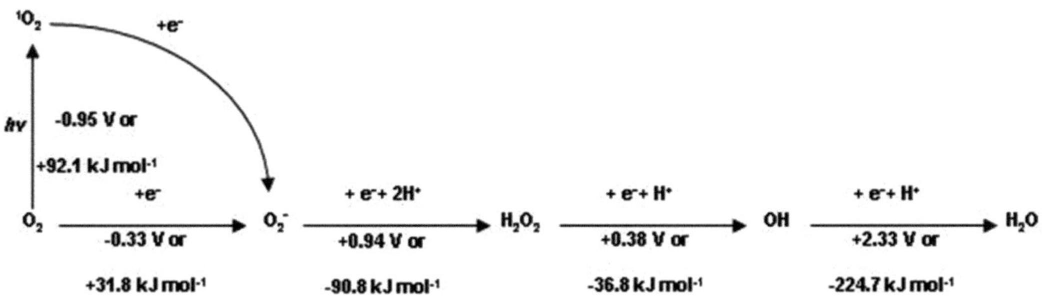

Figure 3. Pathways in the univalent reduction of oxygen to water leading for the formation of various intermediate reactive oxygen species (ROS). Numbers give approximate redox potentials (in volts) or the standard free energy of the reaction (in kJ mol^{-1}).

The superoxide ($^{\bullet}O_2^-$) produced during the first reaction is a short-lived ROS (approximately 2 to 4 μs) and not readily diffusible [72]. In the cellular environment, $^{\bullet}O_2^-$ may cause lipid peroxidation, thereby weakening cell membranes. The second reduction is an exergonic reaction that generates hydrogen peroxide (H_2O_2), a relatively long-lived (1 ms) and stable form of ROS that can diffuse through membranes and therefore reach cellular components distant from its site of synthesis [73]. The last ROS generated by this series of reductions is also exergonic and produces the highly reactive hydroxyl radical ($^{\bullet}OH$), which is the most harmful form of ROS in plant tissues, has a half-life of 1 ϱs and has a very high affinity for biological molecules [74]. The hydroxyl radical is generated from the reaction between $^{\bullet}O_2^-$ and H_2O_2 either spontaneously through the Haber-Weiss reaction or in the presence of reduced transition metals through the Fenton reaction.

Under normal cell conditions, the Haber-Weiss reaction (1) occurs very slowly and very low amounts of $^{\bullet}OH$ are formed:

$$H_2O_2 + {}^{\bullet}O_2^- \rightarrow {}^{\bullet}OH + OH_2 \tag{1}$$

The hydroxyl radical is also formed in very low amounts in the Fenton reaction (2), which is common in biological systems, with its transition metals Fe^{2+} and Cu^+ in a chelated form:

$$H_2O_2 + Fe^{2+} \rightarrow Fe^{3+} + {}^{\bullet}OH + OH \tag{2}$$

The availability of Fe^{2+} limits the reaction rate, but Fe^{3+} can be efficiently reduced by superoxide, thereby maintaining the Fenton reaction ongoing and leading to the generation of $^{\bullet}OH$, as shown in the two half reactions (3) and (4):

$$H_2O_2 + Fe^{2+}\left(Cu^+\right) \rightarrow Fe^{3+}\left(Cu^{2+}\right) + {}^{\bullet}OH + OH \tag{3}$$

$${}^{\bullet}O_2^- + Fe^{3+}\left(Cu^{2+}\right) \rightarrow Fe^{2+}\left(Cu^+\right) + O_2 \tag{4}$$

The prevention of the Haber-Weiss and Fenton reactions is achieved when H_2O_2 and $^{\bullet}O_2^-$ are eliminated prior to these molecules entering into contact with each other.

Due to the high reactivity of $^{\bullet}OH$ radicals, which is the main cause of cell damage under oxidative stress, it is difficult to control their concentration enzymatically. Therefore, plants reduce the presence of this radical by controlling the upstream reactions of $^{\bullet}OH$ formation via Haber-Weiss/Fenton reactions through the elimination of H_2O_2 and $^{\bullet}O_2^-$ prior to their contact with each other. The efficient destruction of $^{\bullet}O_2^-$ and H_2O_2 requires the coordinated action of several antioxidative enzymes and a network of low molecular mass antioxidants.

4.3. Antioxidative system

To mitigate oxidative harm from ROS, plants possess a complex antioxidative system that involves both non-enzymatic and enzymatic antioxidant defences. Non-enzymatic defences include hydrophilic compounds, such as ascorbate and reduced glutathione, and lipophilic compounds, such as tocopherols and carotenoids, which are capable of quenching ROS. Enzymatic defences include superoxide dismutase, catalase and peroxidase. Moreover, an entire array of enzymes is needed for the regeneration of the active forms of antioxidants (glutathione reductase, monodehydroascorbate reductase and dehydroascorbate reductase) [70, 75].

4.3.1. Superoxide Dismutases (SOD)

Superoxide dismutases (EC 1.15.1.1) catalyse the dismutation of superoxide into hydrogen peroxide and water. SOD activity modulates the relative amounts of $^{\bullet}O_2^-$ and H_2O_2 (the two Haber-Weiss reaction substrates) and decreases the risk of the formation of the $^{\bullet}OH$ radical. Since SOD is one of the ubiquitous enzymes in aerobic organisms and is present in most subcellular compartments that generate ROS, this enzyme is considered to play a key role in cell defence mechanisms against ROS [76, 77]. The product of SOD activity is H_2O_2, which is toxic and must be eliminated by conversion into H_2O in subsequent reactions. Although a number of enzymes regulate the intracellular levels of H_2O_2 in plants, catalases and peroxidases are considered to be the most important.

4.3.2. Catalases (CAT)

Catalases (EC 1.11.1.6) are tetrameric heme-containing enzymes that catalyse the dismutation of hydrogen peroxide into water and molecular oxygen, thereby protecting the cell from the harmful effects of H_2O_2 accumulation. CAT is found in all aerobic eukaryotes and is associated with the removal of H_2O_2 generated in biochemical processes, such as the β-oxidation of fatty acids, the glyoxylate cycle (photorespiration) and purine catabolism. CAT activity may decrease under salt stress, heat shock or cold stress, which may be related to plant tolerance to the secondary oxidative stress induced by these forms of environmental stress.

4.3.3. Peroxidases and enzymes regenerating active forms of ascorbate and glutathione

Peroxidases constitute a class of enzymes in the tissues of animals, plants and microorganisms and catalyse the oxidoreduction between hydrogen peroxide and different reductants. There are three classes of plant peroxidases, but ascorbate peroxidase (APX), class III plant peroxidases [or non-specific peroxidases or guaiacol-type peroxidase (POX)] and glutathione peroxidase (GPX) are considered to be the most important plant peroxidases related to the antioxidative system.

Ascorbate peroxidase (EC 1.11.1.11) catalyses the reduction of H_2O_2 to H_2O and has high specificity and affinity for ascorbate (ASC) as a reductant. Its sequence is distinct from other peroxidases and different forms of APX are found in the chloroplasts, cytosol, mitochondria, peroxisomes and glyoxysomes [78]. APX seems to play a key role as a scavenger of H_2O_2 that could leak from these cell organelles.

APX uses two ASC molecules to reduce H_2O_2 to water and produce two monodehydroascorbate (MDHA) molecules (Figure 2). MDHA is a short-lived radical that can either spontaneously dismutate to ascorbate and dehydroascorbate (DHA) (Figure 2) or be reduced to ascorbate by NAD(P)H via monodehydroascorbate reductase (MDHAR; EC 1.6.5.4) (Figure 2), which is found in different cell compartments [16] (Asada, 1997). DHA is reduced to ascorbate by the action of dehydroascorbate reductase (DHAR; EC 1.8.5.1), using reduced glutathione (GSH) as the reducing substrate. This reaction generates reduced glutathione (GSSG), which is, in turn, re-reduced to GSH by NADPH, a reaction catalysed by glutathione reductase (GR; EC 1.6.4.2). The removal of H_2O_2 through this series of reactions is known as the ascorbate-glutathione cycle or the Halliwell-Asada pathway (Figure 2) [75]. Ascorbate and glutathione are not consumed in this pathway, but participate in the cyclic transfer of reducing equivalents, which allows the reduction of H_2O_2 to H_2O, with NADPH as the reducing equivalent donor.

Class III plant peroxidase (EC 1.11.1.7) is a plant-specific oxidoreductase, the activity of which was described as early as 1855. This enzyme is a heme-containing glycoprotein encoded by a large multigene family in plants. POX, which is found in the cytosol, vacuole and cell wall, is less specific to the electron donor substrate than APX and decomposes H_2O_2 through the oxidation of co-substrates, such as phenolic compounds and/or ascorbate [79]. This enzyme is relatively stable at high temperatures and its activity is easily measured using simple chromogenic reactions.

The different types of GPX (EC 1.11.1.9) form a large family of diverse isozymes that reduce H_2O_2 and organic and lipid hydroperoxides using GSH as a reducing agent. In plants, however, it has been suggested that GPX preferably uses thioredoxin as a reductant [80, 81]. Most cellular GPXs are tetrameric enzymes with four identical 22 kDa subunits, each containing a selenocysteine residue in the active site [82]. Selenocysteine participates directly in electron donation to the peroxide substrate and becomes oxidised in the process. The enzyme then uses reduced glutathione as a hydrogen donor to regenerate selenocysteine. GPX uses two GSH molecules to reduce H_2O_2 to water and produce a GSSG molecule (Figure 4).

Taken together, the major ROS-scavenging pathways in plants include SOD, found in almost all cell compartments, CAT in peroxisomes, POX in the cytosol, vacuole and cell wall and the ascorbate-glutathione cycle in the chloroplasts, cytosol, mitochondria, apoplast and peroxisomes. As mentioned above, CAT has extremely high maximal catalytic rates, but low substrate affinities, while APX has a much higher affinity for H_2O_2 than CAT. The high affinity of APX for H_2O_2, in conjunction with the finding of the ascorbate-glutathione cycle in nearly all cell compartments, suggests that this cycle plays a crucial role in controlling the level of ROS in these compartments. Moreover, APX might also be responsible for the fine modulation of H_2O_2 for signalling. In contrast, CAT, which is only present in peroxisomes, is indispensable to H_2O_2 detoxification during stress, when high levels of ROS are produced.

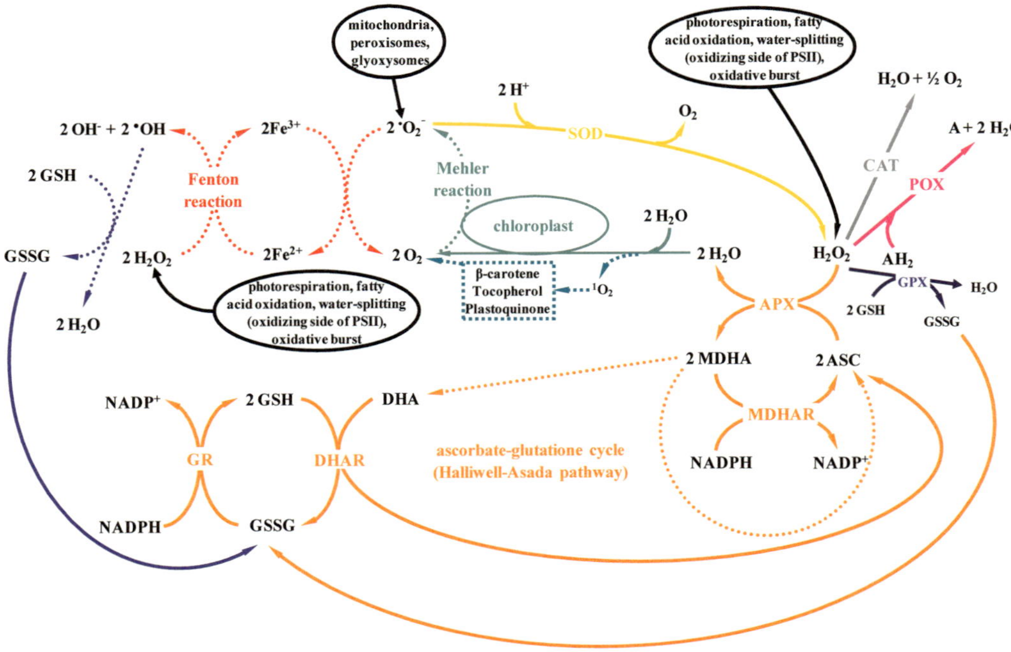

Figure 4. Generation of $^{\bullet}OH$ by Fenton reaction (in red); $^{\bullet}O_2^-$ in the mitochondria, peroxisomes and glyoxysomes and by Mehler reaction in chloroplast (in green), singlet oxygen in chloroplast (in dark green), and H_2O_2 by SOD, photorespiration, fatty acid oxidation or other reactions. SOD acts as the first line of defense converting $^{\bullet}O_2^-$ into H_2O_2 (in yellow). CAT (in grey), POX (in pink), GPX (in dark blue), and APX (in orange) then detoxify H_2O_2. In contrast to CAT, APX requires ASC, POX requires phenolic compounds and/or ASC, and GPX requires GSH as electron donor substrate. In the removal of H_2O_2 through the ascorbate-glutathione cycle (in orange), ASC and GSH participate of the cyclic transfer of reducing equivalents. This cycle uses NADPH as reducing power. $^{\bullet}OH$ may be removed by GSH (in blue), and the GSSG formed is regenerated via GR. Although the pathways of generation and scavenging in the different cell compartments are separate, H_2O_2 can easily diffuse through membranes and antioxidants such as GSH and ASC can be transported between the different compartments. Non-enzymatic pathways are indicated by dotted lines. Abbreviations: APX, ascorbate peroxidase; ASC, ascorbate; AH2, oxidizable substrate; DHA, dehydroascorbate; DHAR, dehydroascorbate reductase; GPX, glutathione peroxidase; POX, non-specific peroxidase; GR, glutathione reductase; GSH, reduced glutathione; GSSG, oxidized glutathione; hydrogen peroxide (H_2O_2); hydroxyl radical ($^{\bullet}OH$); MDHA, monodehydroascorbate; MDHAR, monodehydroascorbate reductase SOD, superoxide dismutase; superoxide radical ($^{\bullet}O_2^-$).

4.4. ROS production and scavenging in drought-stressed plants

The root system is the first plant organ to detect a reduction in the water supply. Besides water and minerals, the roots send signals to the shoots through the xylem sap and the phytohormone abscisic acid is considered to be one of the major root-to-shoot stress signals [83]. In leaves, abscisic acid triggers stomatal closure and the plant shifts to a water-saving behaviour. By controlling the stomatal opening, plants reduce water loss by decreasing the transpiration flux. However, the entrance of carbon dioxide (CO_2) is also reduced simultaneously. This plant response has direct and indirect effects on the net photosynthesis and overall production of ROS under water deficit conditions [84]. A number of studies report increased ROS accumulation and oxidative stress in plants under drought stress [85, 86]. When stomata close in order to limit water loss, there is the occurrence of either a restricted CO_2 supply or CO_2-limited carbon fixation and reduced $NADP^+$ regeneration through the Calvin cycle. Photosynthetic electron transport is, however, maintained at a relatively higher rate in the stressed leaves in comparison to the accentuated reduction in the CO_2 fixation rate [87]. This imbalance between the electron transport and CO_2 fixation rates results in an accentuated reduction of the electron transport chain and the transfer of electrons to O_2 through the Mehler reaction [88]. One study estimated a 50% increase in the leakage of photosynthetic electrons through the Mehler reaction in drought-stressed wheat plants in comparison to non-stressed plants [89].

The photorespiratory pathway is also enhanced under drought stress, especially when the oxygenation of ribulose-1,5-bisphosphate is maximal due to limited CO_2 fixation [90]. Thus, O_2-dependent electron flow and photorespiration can be considered common mechanisms that plants employ to protect the photosynthetic electron transport chain components from photodamage during water deficit. Although it is very difficult to discriminate the amount of ROS generated by the Mehler reaction from that generated by photorespiration, it has been estimated that photorespiration is likely to account for over 70% of total H_2O_2 production under drought stress conditions [90]. In such a scenario, there is considerable potential for the increased accumulation of $^{\bullet}O_2^-$ and H_2O_2 in plants [91]. In a number of plant species, an increased formation of ROS, lipid peroxidation and protein modification have been observed under water deficit conditions [92-94]. The following the sequence of events occurs in plant tissues subjected to such conditions: 1) increased production of ROS and oxidised target molecules; 2) increased expression of genes for antioxidant functions; and 3) increased the levels of non-enzymatic and enzymatic antioxidants, resulting in tolerance to drought stress [95].

Drought stress enhances the *de novo* synthesis of some antioxidative enzymes to overcome the increase in oxidative stress. In rice plants, the *de novo* synthesis of MDHAR, DHAR and GR increases the capacity for ASC and GSH regeneration, which is considered to be one of the primary responses to water deficit so as to mitigate oxidative stress [92, 93]. An increase in the activity of antioxidative enzymes has been reported in a number of plant species submitted to drought stress, enhancing the capacity of the antioxidative system to scavenge ROS and thereby suppressing the level of lipid peroxidation under drought conditions [93, 96, 97].

Additionally, the increase in the activity of antioxidative enzymes and antioxidant content under water deficit conditions appears to be extremely variable among different plant species and even cultivars of the same species. Thus, comparative studies using drought-tolerant and drought-sensitive genotypes demonstrate greater antioxidant capacity in tolerant genotypes. In one study, among five mulberry cultivars subjected to drought, two had efficient antioxidative characteristics that could provide better protection against oxidative stress in leaves under water-limited conditions [98]. Under water stress, a drought-tolerant maize genotype exhibited lower MDA and H_2O_2 contents and an increase in the SOD, CAT, and POX activities in comparison to a drought-sensitive maize genotype [99]. A drought-tolerant wheat genotype exhibited greater APX and CAT activities, higher ASC content and lower H_2O_2 and MDA contents in comparison to a drought-susceptible wheat genotype

[100]. In response to water deficit, the drought-sensitive apple rootstock *Malus hupehensis* exhibited greater increases in H_2O_2, $^{\bullet}O_2^-$ and MDA levels than the drought-tolerant *M. prunifolia*. In contrast, SOD, POX, APX, GR and DHAR activities and ASC and GSH contents increased to a greater extent in *M. prunifolia* than *M. hupehensis* [101]. It has also been reported that the drought-acclimated leaves of wheat plants exhibited a systematic increase in the APX and CAT activities and the maintenance of an adequate ascorbate redox pool through the efficient functioning of the APX enzyme. As a result, lesser membrane damage was found in the drought-acclimated plants [94, 102].

The drought response of a plant species also depends on the duration and severity of the drought period. SOD and CAT activities are reported to have increased in response to severe water deficit in mature leaves of two clones of *Populus deltoids* x *nigra* [103]. For both clones, Mn-SOD, Fe-SOD, and Cu/Zn-SOD isoforms were detected in varying amounts, depending on drought intensity.

Taken together, these findings provide additional evidence that the antioxidative system plays a key role in the process of plant acclimation to drought stress. Thus, greater protection from drought-induced oxidative damage may, at least in part, be involved in tolerance to water deficit.

5. Drought and ecosystems: changes in natural cycles and functional groups

According to Chapman [104], there are an estimated 390,800 plant species worldwide (Magnoliophyta, gymnosperms, ferns, allies and Bryophyta). Despite their occurrence on all continents, biodiversity and distribution is quite variable even within a few kilometres. From an ecologic standpoint, the occurrence of a specific plant species in an area depends on the combination of three factors:

a. **Chance** – the possibility of a propagule reaching and establishing itself in a certain location;

b. **History** – the current abundance of a species is probably correlated with its abundance in the near past;

c. **Necessity** – demands for growth, competence for competition and interactions with other organisms; Coexistence with other plants depends on the complexity of the environment in terms of fertility, sunlight and water availability and on how strongly the plant can withstand the action of competitors, herbivores, parasites, etc.

Among these needs, water availability can be considered the most influential and even shapes the phytophysiognomy of some ecosystems. According to Puig [105], while drought has little influence in a tropical rain forest (where precipitation surpasses evapotranspiration more than ten months per year), water regime variability in a tropical dry forest is the major

http://dx.doi.org/10.5772/53833

environmental factor exerting an influence on the ecological processes that regulate its vegetation maintenance and distribution [106].

On the ecosystem level, a drought event can be (i) **permanent** – in regions where a desert climate predominates; (ii) **seasonal** – as observed in semi-arid regions; (iii) **irregular or variable** – as occurs in regions with humid or sub-humid climates (this normally takes place in limited areas and the return of drought is unpredictable); or (iv) **invisible or green drought** – as occurs when precipitation is not interrupted, but lesser than evapotranspiration, causing a regional moisture imbalance. In the latter case, there is a drop in relative air humidity, leading to a reduction in moisture content in the soil. Moisture is evaporated into the atmosphere and comes back as rainfall, but not enough to increase the moisture content in the soil. This is considered the worst kind of drought due to the fact that is difficult to perceive.

5.1. Formation of functional groups under natural cycles

Excessive insolation, fire, shade, wind, herbivory, nutrient availability and water availability are factors that force plants to exhibit different kinds of adaptation to overcome the constraints to their survival and establishment. In some cases, plant species from unrelated taxonomical groups use very similar strategies, resulting in a phenomenon denominated convergent evolution.

Cummins [107] proposed a plant classification system based on similar roles or analogous processes in the ecosystem. This classification allows us to simplify the biodiversity in a given location and correlate it with that of another location, even without taxonomic relatedness among the species found [108]. A number of papers have since been published revealing the existence of vegetation patterns as responses to the influence of biotic and/or abiotic factors in different ecosystems. Consequently, knowledge on how an assemblage of plants organises itself to occupy all available niches under given environmental conditions has continually increased. The three general mechanisms used by plants to cope with drought [**avoidance** (dormancy in the dry season), **delay** (through increased water uptake and reduced water loss) and **physiological tolerance** (maintenance of plant functioning with low cell water content)] are closely linked to the functional traits of the species [109] (Table 2).

Functional trait	Role	Some co-existing species	Source
Life form	Species can avoid drought remaining as seed during dry season (Therophytes)	*Gomphrena* aff. *leucocarpa* Mart (Amaranthaceae) *Taccarum peregrinum* L (Araceae) *Pithecoseris pacourinoides* Mart (Asteraceae) *Cleome guianensis* Aublet (Capparaceae) *Euphorbia comosa* Vell. (Euphorbiaceae) *Cuphea ericoides* Cham. & Schlech (Lythraceae) *Richardia scabra* L. (Rubiaceae) *Amasonia campestris* L. (Verbenaceae)	Mendes [110]

Functional trait	Role	Some co-existing species	Source
Specific leaf area	This is an index of sclerophylly.	*Prunus ilicifolia* (Nutt. ex Hook. & Arn.) Walp. (Rosaceae) *Ceanothus oliganthus* var. *sorediatus* (Rhamnaceae) *Mimulus aurantiacus* Curtis (Phrymaceae) *Baccharis pilularis* DC. (Asteraceae)	Ackerly et al. [111]
Leaf size	Influences leaf cooling and light capture efficiency (self-shading)	*Cercocarpus betuloides* Nutt. (Rosaceae) *Comarostaphylis diversifolia* (Parry) Greene (Ericaceae) *Quercus agrifolia* Née (Fagaceae)	Scoffoni et al. [112]
Leaf phenology	Plays an important role in drought resistance, as deciduous trees are able to reduce water loss by dropping leaves, while evergreen trees must resist drought.	EVERGREEN *Capparis flexuosa* L. (Capparaceae) *Maytenus rigida* Mart. (Celastraceae) *Licania rigida* Benth. (Chysobalanaceae) *Ximenia americana* L. (Olacaceae) DECIDUOUS *Amburana cearensis* (Allemão)AC Smith (Faboideae) *Jatropha mollissima*(Pohl) Baill.(Euphorbiaceae) *Combretum leprosum* Mart. (Combretaceae) Pseudobombax marginatum (A. St. –Hil.,Juss&Camb.) A. Robyns (Bombacaceae)	Barbosa et al. [113]
Stem / Wood density (WD)	This is negatively correlated with cavitation resistance and negatively correlated with water storage.	*Anogeissus latifolia* (Roxb. Ex DC) Wall. ex Bedd. (Combretaceae) *Soymida febrifuga* (Roxb.) A. Juss. (Meliaceae) *Acacia catechu* (L. f.) Willd. (Fabaceae) *Shorea robusta* Roth (Dipterocarpaceae) *Chloroxylon swietenia* DC. (Rutaceae)	Kushwaha et al. [114]
Root deep	Allows an exploration of the moister deeper soil layers	SHALLOW *Schefflera macrocarpa* (Seem.) D. C. Frodin (Araliaceae) *Miconia ferruginata* DC. (Melastomataceae) *Roupala Montana* Aubl. (Proteaceae) *Ouratea hexasperma* (St. Hil.) Baill. (Ochnaceae) DEEP *Vochysia elliptica* Mart. (Vochysiaceae) *Dalbergia miscolobium* Benth. (Fabaceae) *Kielmeyera coriacea* Mart. (Clusiaceae)	Franco et al. [115]

Table 2. Some functional traits associated to drought tolerance in plants under dry conditions.

http://dx.doi.org/10.5772/53833

5.2. Climate change: New challenge for plants

Drought is a deviation from normal climatic conditions in which there is a lack of precipitation over an extended period and the resulting water shortage has negative implications [116]. Drought differs from aridity, which is a normal condition of a severe lack of water availability in a specific region.

In recent decades, the planet has witnessed intense climate changes due to global warming. Extreme climatic events, such as tornados, hurricanes, floods, blizzards and drought, have become more frequent and intense. Some annual plant events, such as flowering, fruiting and re-sprouting, follow a specific timing, which is denominated phenology. Global warming can affect this timing and its consequences can affect water supplies, pollination and the overall functioning of natural and agricultural ecosystems. This situation suggests a bleak future for mankind and nature, as all organisms will face substantial disturbances in their environment, possibly beyond their capacity for resistance and resilience. Resistance is the ability of a system to maintain its structure and functioning after a disturbance and resilience is the ability to re-establish equilibrium after it has been disrupted [117].

A given plant species can either escape from or acclimate to adverse environmental conditions, which can change in space and time. When a specific genotype exteriorises different phenotypes under different conditions, it is considered to have adequate phenotypic plasticity. Changes in the partitioning of resources can be the result of different strategies under different selection pressures. However, this phenotypic plasticity is quite limited due both the physiological costs and ontogenetic drift [118, 119].

The following are the most detectable features of global warming: 1) its influence on the perception of plants regarding the seasons (the advance of biological spring and the delay in biological winter have been observed and such changes have a direct effect on the reproductive events of flowering and fructification, which can affect the dynamics of plant populations and communities) [120-122]; 2) alterations in the floristic composition and phytosociology of plant communities due to changes in the seedling mortality rate; 3) the occurrence of a climate-induced shift in the range of species, which can force the interaction of plants with those from which they were formerly spatially separated [123]; and 4) increased biological plant invasions, as global warming can modify the dynamics and climate of new environments, making them suitable for invasion [124, 125].

Despite the volume of studies on plant responses to global warming, a great deal of uncertainty remains. After an extensive survey of plant phenology databases for long-term observations and short-term warming experiments involving 1634 species, Wolkovich et al. [126] concluded that such experimental studies underpredict plant phenological responses to global warming. Thus, more in-depth studies are needed to help predict the effects of global warming on plant communities in the near future and develop strategies to mitigate these effects.

Author details

Elizamar Ciríaco da Silva[1*], Manoel Bandeira de Albuquerque[2], André Dias de Azevedo Neto[3] and Carlos Dias da Silva Junior[1]

*Address all correspondence to: elizaciriaco@gmail.com

1 Laboratory of Applied Botany, Department of Biology, Federal University of Sergipe, Brazil

2 Laboratory of Plant Ecology, Department of Crop and Environmental Sciences, Federal University of Paraíba, Brazil

3 Laboratory of Biochemistry, Center of Exate and Technological Sciences, Federal University of Recôncavo da Bahia, Brazil

References

[1] Agbola T, Ojeleye D. Climate change and food crop production in Ibadan, Nigeria. African Crop Science Conference Proceedings 2007; 8:1423-1433.

[2] Reynolds MP, Ortiz R. Adapting crops to climate changes: a summary. In: Reynolds MP (ed.) Climate Change and Crop Production. CABI series in climate change v.1. Chippenam: CPI; 2010. p1-8.

[3] Silva EC, Nogueira RJMC, Silva MA, Albuquerque MB. Drought stress and plant nutrition. Plant Stress 2011; 5(1): 32-41.

[4] Assad ED, Pinto HS, Zullo Junior J, Ávila AMH. Impacto das mudanças climáticas no zoneamento agroclimático do café no Brasil. Pesquisa Agropecuária Brasileira 2004; 39(11):1057-64.

[5] Abraham EM, Beekman GB. Indicadores de la desertificación para América Del Sur. Editorial Martin Fierro: Mendonza; 2006.

[6] Primack R, Rodrigues E. Biologia da conservação. Londrina: E. Rodrigues; 2001.

[7] Morales C. Pobreza, desertificación y degradación de tierras. In: Morales C, Parada S (eds.) Pobreza, desertificación y degradación de los recursos naturalis. CEPAL: Santiago de Chile; 2005. p.25-58.

[8] Sarker BC, Hara M, Uemura M. Proline synthesis, physiological responses and biomass yield of eggplants during and after repetitive soil moisture stress. Scientia Horticulturae 2005; 103:387-402.

[9] Sircelj H, Tausz M, Grill D, Batic F. Biochemical responses in leaves of two apple tree cultivars subjected to progressing drought. Journal of Plant Physiology 2005; 162:1308-18.

[10] Silva EC, Nogueira RJMC, Vale FHA, Melo NF, Araujo FP. Water relations and organic solutes production in four umbu tree (Spondias tuberosa) genotypes under intermittent drought. Brazilian Journal of Plant Physiology 2009; 21(1): 43-53.

[11] Taiz L, Zeiger E. Plant Physiology. Fourth Edition. Sinauer Associates: Los Angeles; 2006.

[12] Ajum SA, Xie XY, Wang LC, Saleem MF, Man C, Lei W. Morphological, physiological and biochemical responses of plants to drought stress. African Journal of Agricultural Research 2011; 6(9):2026-32.

[13] Sakamoto A, Murata N. The role of glycinebetaine in the protection of plants from stress: clues from transgenic plants. Plant Cell and Environment 2002; 25:163-171.

[14] Ashraf M, Foolad MR. Roles to glycine betaine and proline in improving plant abiotic stress resistance. Environmental and Experimental Botany 2007; 59: 206-216.

[15] Chaves MM, Maroco JP, Pereira JS. Understanding plant response to drought – from genes to the whole plant. Functional Plant Biology 2003; 30: 239-264.

[16] Jaleel CA, Manivannan P, Wahid A, Farooq M, Al-Juburi HJ, Somasundaram R, Vam RP. Drought stress in plants: a review on morphological characteristics and pigments composition. International Journal of Agriculture and Biology 2009; 11(1):100-105.

[17] DaMatta FM, Ramalho JDC. Impacts of drought and temperature stress on coffee physiology and production: a review. Brazilian Journal of Plant Physiology 2006; 18(1) 55-8.

[18] Saldaña-Zorrilla SO. Socioeconomic vulnerability to natural disasters in Mexico: rural poor, trade and public response. México: CEPAL; 2007.

[19] Franchini JC, Debiasi H, Nepomuceno AL, Farias RB. Manejo do solo para redução das perdas de produtividade por seca. Embrapa, 2009. http://bioinfo.cnpso.embrapa.br/seca/index.php/manejo-do-solo (acessed 10 August 2012).

[20] Chavarria G, Santos HP. Plant Water Relations: Absorption, Transport and Control Mechanisms. In: Montanaro G, Dichio B (eds.) Advances in Selected Plant Physiology Aspects. Rijeka: InTech; 2012. p105-132.

[21] Fitter AH, Hay RKM. Environmental physiology of plants. London: Academic Press; 2002.

[22] Miloud H, Ali G. Some aspects of leaf senescence. In: Nagata T. (ed.) Senescence. Rijeka: InThec; 2012. p107-116.

[23] Pinheiro HA, DaMatta FM, Chaves ARM, Loureiro ME. Drought tolerance is associated with rooting depth and stomatal control of water use in clones of Coffea canephora. Annals of Botany 2005; 96: 101–108.

[24] Gindaba J, Rozanov A, Negash L. Photosynthetic gas exchange, growth and biomass allocation of two Eucalyptus and indigenous tree species of Ethiopia under moisture deficit. Forest Ecology and Management 2005; 205:127-138.

[25] Prado CHBA, Wenhui Z, Rojas MHC, Souza GM. Seasonal leaf gas exchange and water potential in a woody cerrado species community. Brazilian Journal of Plant Physiology 2004; 16(1):7-16.

[26] Hare, PD, Cress WA, Van Staden J. Dissecting the roles of osmolyte accumulation during stress. Plant, Cell and Environment 1998; 21:535-553.

[27] Hong-Bo F, Xiao-Yan C, Li-Ye C, Xi-Ling Z, Gang W, Yong-Bing Y, Chang-Xing Z, Zan-Min H. Investigation on the relationship of proline with wheat anti-drought under soil water deficits. Colloids and Surfaces B 2006; 53:113-119.

[28] Sánchez FJ, Manzanares M, Andres EF, Tenorio JL, Ayerba L. Turgor maintenance, osmotic adjustment and soluble sugar and proline accumulation in 49 pea cultivars in response to water stress. Field Crops Research 1998; 59:225-235.

[29] Rabe ES. Stress physiology: the functional significance of the accumulation of nitrogen-containing compounds. Journal of Horticultural Science 1990; 65(3):231-243.

[30] Knipp G, Honermeier B. Effect of water stress on proline accumulation of genetically modified potatoes (Solanum tuberosum L.) generating fructans. Journal of Plant Physiology 2006; 163:392-397.

[31] Kellomäki S, Strandman H, Nuutinen T, Petola H, Kothonen KT, Väisänen H. Adaptation of forest ecosystems, Forest and Forestry to Climate Change. FINDAT Working Paper 4. Helsinki: Finnish Environment Institute Mimeographs. 2005. 44 p.

[32] Angelopoulos K, Dichio B, Xiloyannis C. Inhibition of photosynthesis in olive trees (Olea europaea L.) during water stress and rewatering. Journal of Experimental Botany 1996; 47:1093-1100.

[33] Nepomuceno AL, Oosterhuis DM, Stewart JM. Physiological responses of cotton leaves and roots to water deficit induced by polyethylene glycol. Environmental and Experimental Botany 1998; 40:29-41.

[34] Eckstein K, Robinson JC. Physiological responses of banana (Musa AAA; Cavendish sub-group) in the subtropics. VI. Seasonal responses of leaf gas exchange to short-term water stress. Journal of Horticultural Science 1996; 71:679-692.

[35] Bolhàr-Nordenkampf HR, Long SP, Baker NR, Öquist G, Schreiber U, Lechner EG. Chlorophyll fluorescence as a probe of the photosynthetic competence of leaves in the field: A review of current instrumentation. Functional Ecology 1989; 3:497-514

[36] Bolhàr-Nordenkampf HR, Öquist G. Chlorophyll fluorescence as a tool in photosynthesis research. In: Hall DO, Scurlock JMO, Bolhàr-Nordenkampf HR, Leegood RC, Long SP (eds) Photosynthesis and Production in a Changing Environment: a Field and Laboratory Manual. London: Chapman & Hall; 1993.

[37] Baker NR. Light-use efficiency and photoinhibition of photosynthesis in plants under environmental stress. In: Smith JAC, Griffiths H (eds) Water deficits plant responses from cell to community. Oxford: Bios Scientific Publisher; 1993. p221-235.

[38] Matoušková, M., Nauš, J., Flašarová, M., Fiala, J. Changes in curves of fast fluorescence induction caused by water stress of barley plants. – Acta Univ. Palacki. Olomouc, Fac. Rer. Nat. Physica 1996; 35:195-208.

[39] Berry J, Björkman O. Photosynthetic response and adaptation to temperature in higher plants. Annual Review of Plant Physiology 1980; 31:491-543.

[40] Pimentel C. Metabolismo de carbono na agricultura tropical. Seropédica:EDUR;1998.

[41] Gilmore AM, Govindjee. How higher plants respond to excess light: energy dissipation in photosystem II. In: Singhal Gs, Renger G, Sopory SK, Irrgang KD, Govindjee (eds), Concepts in photobiology: photosynthesis and photomorphogenesis. New Delhi: Narosa Publ.; 1999. p513-548.

[42] Raison JK, Roberts JKM, Berry JA. Correlations between thermal stability of chloroplast (thylakoid) membranes and the composition and fluidity of their polar lipids upon acclimation of the higher plant Nerium oleander, to growth temperature. Biochimica et Biophysica Acta 1982; 688:218-228.

[43] Yamane Y, Kashino Y, Koike H, Satoh K. Effects of high temperatures on the photosynthetic systems in spinach: oxygen-evolving activities, fluorescence characteristics and the denaturation process. Photosynthesis Research 1998; 57:51-59.

[44] Pastenes C, Horton P Effect of high temperature on photosynthesis in beans. II. CO_2 assimilation and metabolite contents. Plant Physiology 1996; 112:1253-1260.

[45] Pastenes C, Horton P. Resistance of photosynthesis to high temperature in two bean varieties (Phaseolus vulgaris L.) Photosynthesis Research 1996; 62: 197-203.

[46] Briantais JM, Dacosta J, Goulas Y, Ducruet JM, Moya I, Heat stress induces in leaves an increase of the minimum level of chlorophyll fluorescence, F0 : a time-resolved analysis. Photosynthesis Research 1996; 48:189-196.

[47] Havaux M, Ernez M, Lannoye R. Correlation between heat tolerance and drought tolerance in cerals demonstrated by rapid chlorophyll fluorescence. Journal of Plant Physiology 1998; 133: 555-560.

[48] Yamane Y, Kashino Y, Koile H, Satoh K.Increase in the fluorescence F0 level reversible inhibition of Photosystem II reaction center by high-temperature treatments in higher plants. Photosynthesis Research 1997; 52:57-64.

[49] Maxwell K, Johnson GN. Chlorophyll fluorescence - A pratical guide. Journal of Experimental Botany 2000;51: 659-668.

[50] Saddi, N. A taxonomic revision of the genus Kielmeyera Mart. (Guttiferae). PhD thesis, University of Reading, UK; 1982.

[51] Alves TMA, Silva AF, Brandão M, Grandi, TSM, Smania, EF, Smania Junior, AS, Zani, CL. Biological screening of Brazilian medicinal plants. Memorial do Instituto Oswaldo Cruz 2000; 95: 367–373.

[52] Zouni A, Witt HT, Kern J, Fromme P, Kraub N, Saenger W, Orth P. Crystal structure of hotosystem II from Synecococcus elongatesat 3.8 A resolution. Nature 2001; 409:739-43.

[53] Strasser, RJ, Tsimilli-Michael M, Srivastava A. Analysis of the chlorophyll a fluorescence transient. In Papageorgiou GC, Govindjee (eds.) Chlorophyll a fluorescence: A signature of photosynthesis. Dordrecht: Springer; 2004. p321–62.

[54] Strasser RJ. The grouping model of plant photosynthesis: heterogeneity of photosynthetic units in thylakoids. In: Akoyunoglou G (ed.). Photosynthesis III. Structure and molecular organisation of the photosynthetic apparatus. Philadelphia: Balaban International Science Services; 1981. p727-737.

[55] Tsimilli-Michael M, Strasser R. In vivo assessment of stress impact on plants' vitality: applications in detecting and evaluating the beneficial role of Mycorrhization on host plants. In: Varma A (ed.) Mycorrhiza: State of the art, genetics and molecular biology, ecofunction, biotechnology, eco-physiology, structure and systematics. Berlin: Springer; 2008. p679-703.

[56] Krause GH, Weis E. Chlorophyll fluorescence as a tool in plant physiology. II. Interpretation of fluorescence signals. Photosynthesis Research 1984; 5:139-157.

[57] Bulkhov N, Wiese C, Neimanis S, Heber U. Heat sensitivity of chloroplasts and leaves: Leakage of protons from thylakoids and reversible activation of cyclic eletron transport. Photosynthesis Research 1999; 59:81-93.

[58] Andréasson LE, Vass I, Styring S. Ca^{2+} depletion modifies the electron transfer on the both donor and acceptor sides in photosystem II from spinach. Biochimica et Biophysica Acta 1995; 1230:155-164.

[59] Enami I, Kitamura M, Tomo T, Isokawa Y, Ohta H, Katch S. Is the primary cause of thermal inactivation of oxygen evolution in spinach PS II membranes release of the extrinsic 33 kDa protein or of Mn? Biochimica et Biophysica Acta 1994; 1186:52-58.

[60] Joliot P, Joliot A. Cyclic electron transport in plant leaf. Proceedings of the National Academy of Sciences of the United States of America 2002; 99:10209-10214.

[61] Strasser RJ, Srivastava A, Tsimilli-Michael M. The fluorescence transient as a tool to characterize and screen photosynthetic samples. In: Yunus M, Pathre U, Mohanty P

(eds), Probing Photosynthesis: Mechanisms, Regulation and Adaptation. London: Taylor & Francis; 2000. p445-483.

[62] Mehta P, Jajoo A, Mathur S, Bharti S. Chlorophyll a fluorescence study effects of high salt stress on photosystem II in wheat leaves. Plant Physiology and Biochemistry 2010; 48: 16-20.

[63] Srivastava A, Strasser RJ, Govindjee. Greening of peas: parallel measurements of 77 K emission spectra, O-J-I-P chlorophyll a fluorescence transient, period four oscillation of the initial fluorescence level, delayed light emission, and P700. Photosynthetica 1999; 37:365-392.

[64] Gonçalves, JFC, Silva, CEM, Guimarâes, DG. Fotossíntese e potencial hídrico foliar de plantas jovens de andiroba. Pesquisa Agropecuária Brasildeira 2009; 44(1):8-14.

[65] Lazar D. Chlorophyll a fluorescence rise induced by high light illumination of dark-adapted plant tissue studied by means of photosystem II and considering photosystem II heterogeneity. Journal of Theoretical Biology 2003; 220:469-503.

[66] Hermans C, Smeyers M, Rodriguez RM, Eyletters M, Strasser RJ, Delhaye JP. Quality assessment of urban trees: A comparative study of physiological characterization, airborne imaging and on site fluorescence monitoring by the O-J-I-P-test. Journal of Plant Physiology 2003. 160:81-90.

[67] Force L, Critchley C, Rensen JJS. New fluorescence parameters for monitoring photosynthesis in plants. Photosynthesis Research 2003; 78:17-33.

[68] Strauss AJ, Kruger GHJ, Strasser RJ, Van Heerden PDR. Ranking of dark chilling tolerance in soybean genotypes probed by chlorophyll a fluorescence transient O-J-I-P. Environmental and Experimental Botany 2006; 56:147-157.

[69] Scandalios J.G. The rise of ROS. Trends in Biochemical Science 2002; 27:483-486.

[70] Azevedo Neto AD, Gomes-Filho E, Prisco JT. Salinity and oxidative stress. In: Khan NA, Singh S (ed.) Abiotic stress and plant responses. New Delhi: I.K. International; 2008. p57-82.

[71] Gill SS, Tuteja N. Reactive oxygen species and antioxidant machinery in abiotic stress tolerance in crop plants. Plant Physiology and Biochemistry 2010; 48(12):909-930.

[72] Smirnoff N. The role of active oxygen in the response of plants to water deficit and desiccation. New Phytologist 1993; 125:27-58.

[73] Willekens H, Chamnongpol S, Davey M, Schraudner M, Langebartels C, Van Montagu M, Inzé D, Van Camp W. Catalase is a sink for H_2O_2 and is indispensable for stress defense in C3 plants. EMBO Journal 1997;16:4806-4816.

[74] Dat J, Vandenabeele S, Vranová E, Van Montagu M, Inzé D, Van Breusegen F. Dual action of the active oxygen species during plant stress responses. Cellular and Molecular Life Sciences 2000; 57:779-795.

[75] McKersie BD. Oxidative stress. In: McKersie B.D., Leshem Y.Y. (eds.) Stress and stress coping in cultivated plants. Dordrecht: Kluwer Academic Publishes; 1994. p15-55.

[76] Bowler C, Van Montagu M, Inzé D. Superoxide dismutase and stress tolerance. Annual Review of Plant Physiology and Plant Molecular Biology 1992; 43:83-116.

[77] Scandalios JG. Oxygen stress and superoxide dismutase. Plant Physiology 1993;101:7-12.

[78] Shigeoka S, Ishikawa T, Tamoi M, Miyagawa Y, Takeda T, Yabuta Y, Yoshimura K. Regulation and function of ascorbate peroxidase isoenzymes. Journal of Experimental Botany 2002; 53:1305-1319.

[79] Hiraga S, Sasaki K, Ito H, Ohashi Y, Matsui H. A. large family of class III plant peroxidases. Plant and Cell Physiology 2001; 42:462-468.

[80] Herbette P, Lenne C, Leblanc N, Julien JL, JoeDrevet R, Roeckel-Drevet P. Two GPX-like proteins from Lycopersicon esculentum and Helianthus annuus are antioxidant enzymes with phospholipid hydroperoxide glutathione peroxidase and thioredoxin peroxidase activities. European Journal of Biological Chemistry 2002; 269:2414-2420.

[81] Jung BG, Lee KO, Lee SS, Chi YH, Jang HH, Kang SS, Lee K, Lim D, Yoon SC, Yun DJ, Inoue Y, Cho MJ, Lee SY. A chinese cabbage cDNA with high sequence identity to phospholipid hydroperoxide glutathione peroxidases encodes a novel isoform of thioredoxin-dependent peroxidase. The Journal of Biological Chemistry 2002; 277:12572-12578.

[82] Arthur JR. The glutathione peroxidases. CMLS, Cellular and Molecular Life Sciences 2000; 57:1825–1835.

[83] Davies W.J., Zhang J. Root signals and the regulation of growth and development of plants in drying soil. Annual Review of Plant Physiology and Plant Molecular Biology 1991;42 55-76.

[84] Mittler R. Oxidative stress, antioxidants and stress tolerance. Trends in Plant Science 2002; 7:405-410.

[85] Sgherri CLM, Pinzino C, Navari-Izzo F. Chemical changes and O_2- production in thylakoid membranes under water stress. Physiologia Plantarum 1993; 87:211-216.

[86] Beis A, Patakas A. Relative contribution of photoprotection and anti-oxidative mechanisms to differential drought adaptation ability in grapevines. Environmental and Experimental Botany 2012; 78:173-183.

[87] Foyer CH, Noctor G. Oxygen processing in photosynthesis: regulation and signaling. New Phytologist 2000; 146(3):359-388.

[88] Asada K. The water-water cycle in chloroplasts: scavenging of active oxygens and dissipation of excess photons. Annual Review of Plant Biology 1999; 50:601-639.

[89] Biehler K, Fock H. Evidence for the contribution of the Mehler-Peroxidase reaction in dissipating excess electrons in drought-stressed wheat. Plant Physiology 1996; 112:265-72.

[90] Noctor G, Veljovic-Jovanovic S, Driscoll S, Novitskaya L, Foyer C. Drought and oxidative load in the leaves of C3 plants: A predominant role for photorespiration? Annals of Botany 2002; 89:841-850.

[91] Robinson JM, Bunce J. A. Influence of drought-induced water stress on soybean and spinach leaf ascorbate-dehydroascorbate level and redox status. International Journal of Plant Sciences 2000; 161:271-279.

[92] Boo YC, Jung J. Water deficit-induced oxidative stress and antioxidant defenses in rice plants. Journal of Plant Physiology 1999; 155:255-261.

[93] Sharma P, Dubey RS. Drought induces oxidative stress and enhances the activities of antioxidant enzymes in growing rice seedlings. Plant Growth Regulation 2005; 46:209-221.

[94] Esfandiari E, Shakiba MR, Mahboob SA, Alyari H, Shahabivand S. The effect of water stress on the antioxidant content, protective enzyme activities, proline content and lipid peroxidation in wheat seedling. Pakistan Journal of Biological Sciences 2008; 11:1916-1922.

[95] Mano J, Torii Y, Hayashi S, Takimoto K, Matsui K, Nakamura K, Inzé I, Babiychuk E, Kushnir S, Asada A. The NADPH: Quinone oxidoreductase P1-ζ-crystallin in Arabidopsis catalyzesthe α, β-hydrogenation of 2-alkenals: Detoxication of the lipid peroxide-derived reactive aldehydes. Plant and Cell Physiology 2002; 43(12):1445-1455.

[96] Azevedo Neto AD, Nogueira RJMC, Melo Filho PA, Santos R. Physiological and biochemical responses of peanut genotypes to water deficit. Journal of Plant Interactions 2010;5 1-10.

[97] Sayfzadeh S, Rashidi M. Response of antioxidant enzymes activities of sugar beet to drought stress. ARPN Journal of Agricultural and Biological Science 2011; 6(4):27-33.

[98] Reddy AR, Chaitanya KV, Jutur PP, Sumithra K. Differential antioxidative responses to water stress among five mulberry (Morusalba L.) cultivars. Environmental and Experimental Botany 2004; 52:33-42.

[99] Moussa HR, Abdel-Aziz SM. Comparative response of drought tolerant and drought sensitive maize genotypes to water stress. Australian Journal of Crop Science 2008; 1:31-36.

[100] Sairam RK, Deshmukh PS, Saxena DC. Role of antioxidant systems in wheat genotypes tolerance to water stress. Biologia Plantarum 1998; 41(3):387-394.

[101] Wang S, Liang D, Li C, Hao Y, Ma F, Shu H. Influence of drought stress on the cellular ultrastructure and antioxidant system in leaves of drought-tolerant and drought-sensitive apple rootstocks. Plant Physiology and Biochemistry 2012; 51:81-89.

[102] Al-Ghamdi AA. Evaluation of oxidative stress tolerance in two wheat (Triticum aestivum) cultivars in response to drought. International Journal of Agriculture and Biology 2009; 11:7-12.

[103] Marron N, Maury S, Rinaldi C, Brignolas F. Impact of drought and leaf development stage on enzymatic antioxidant system of two Populus deltoids x nigra clones. Annals of Forest Science 2006; 63:323-327.

[104] Chapman AD. Numbers of living species in Australia and the World. 2nd edition. Canberra: Australian Government, Department of the Environment, Water, Heritage and the Arts; 2009.

[105] Puig H. A floresta tropical úmida. São Paulo: Editora UNESP; 2008.

[106] Murphy PG, Lugo AE. Ecology of tropical dry forest. Annual Review of Ecology and Systematics 1986; 17:67-88.

[107] Cummins K. Structure and function of stream ecosystems. Bioscience 1974; 24:631-641.

[108] Alvarez-Añorve M, Quesada M, De La Barrera E. Remote sensing and plant functional groups detection: physiology, ecology and spectroscopy in tropical systems. In: Kalacska M, Sánchez-Azofeifa GA (eds.) Hyperspectral Remote Sensing of Tropical and Sub-Tropical Forests. London:Taylor and Francis Group; 2008. p27–45.

[109] Poorter L, Markesteijn L. Seedling traits determine drought tolerance of tropical tree species. Biotropica 2008; 40:321–331.

[110] Mendes MRA. Florística e fitossociologia de um fragmento de caatinga arbórea, São José do Piauí, Piauí. MSc thesis. Universidade Federal de Pernambuco; 2003.

[111] Ackerly DD, Knight CA, Weiss SB, Barton K, Starmer KP. Leaf size, specific leaf area and microhabitat distribution of chaparral woody plants: contrasting patterns in species level and community level analyses 2003. Oecologia; 130:449-457.

[112] Scoffoni C, Rawls M, McKown A, Cochard H, Sack L. Decline of leaf hydraulic conductance with dehydration: Relationship to leaf size and venation architecture. Plant Physiology 2011; 156:832–843.

[113] Barbosa DCA, Barbosa MCA, Lima LCM. Fenologia de espécies lenhosas da caatinga. In: Leal IR, Tabarelli M, Silva JMC (eds.) Ecologia e Conservação da Caatinga. Recife: Editora Universitária da UFPE; 2003. p657-693.

[114] Kushwaha CP, Tripathi SK, Singh GS, Singh KP. Diversity of deciduousness and phenological traits of key Indian dry tropical forest trees. Annals of Forest Science 2010; 67: Article 310. DOI: 10.1051/forest/2009116.

[115] Franco AC, Bustamante M, Caldas LS, Goldstein G, Meinzer FC, Kozovits AR, Rundel P, Coradin VTR. Leaf functional traits of Neotropical savanna trees in relation to seasonal water deficit. Trees 2005; 19:326-335

[116] Wilhite DA, Glantz MH. Understanding the drought phenomenon: The role of definitions. Water International 1985; 10(3):111–120.

[117] Molles Junior MC. Ecology: Concepts and applications. Boston: McGraw-Hill. 2002.

[118] Valladares F, Gianoli E, Goéz JM. Ecological limits to plant phenotypic plasticity. New Phytologist 2007; 176:749-763.

[119] Weiner J. Allocation, plasticity and allometry in plants. Perspectives in Plant Ecology, Evolution and Systematics 2004. 6(4): 207-215.

[120] Intergovernmental Panel on Climate Change. IPCC. The Physical Science Basis: Contribution of Working Group I to the Fourth Assessment of the Intergovernmental Panel on Climate Change. Cambridge: Cambridge university; 2007.

[121] Peñuelas J, Rutishauser T, Filella I. Phenology feedbacks on climate change. Science 2009; 324:887-888.

[122] Sherry RA, Zhou XH, Gu SL, Arnone JA, Schimel DS, Verburg PS, Wallace LL, Luo YQ. Divergence of reproductive phenology under climate warming. Proceedings of the National Academy of Sciences of the United States of America 2007; 104(1): 198-202.

[123] Walther GR. Community and ecosystem responses to recent climate change. Philosophical Transactions of Royal Society B 2010; 365:2019–2024

[124] Bradley BA, Blumenthal DM, Wilcove DS, Ziska LH. Predicting plant invasions in an era of global change. Trends in Ecology and Evolution 2010; 25(5):310-318.

[125] Vila M, Corbin JD, Dukes JS, Pino J, Smith SD. Linking plant invasions to global environmental change. In: Canadell J, Pataki D, Pitelka L (Eds.) Terrestrial Ecosystems in a Changing World, New York: Springer; 2007. p.93-102.

[126] Wolkovich EM, Cook BI, Allen JM, Crimmins TM, Betancourt JL, Travers SE, Pau S, Regetz J, Davies TJ, Kraft NJB, Ault TR, Bolmgren K, Mazer SJ, McCabe GJ, McGill BJ, Parmesan C, Salamin N, Schwartz MD, Cleland EE. Warming experiments underpredict plant phenological responses to climate change. Nature 2012; 485:494-497.

Chapter 3

Tolerance to Drought in Leguminous Plants Mediated by *Rhizobium and Bradyrhizobium*

Allan Klynger da Silva Lobato,
Joaquim Albenísio Gomes da Silveira,
Roberto Cezar Lobo da Costa and
Cândido Ferreira de Oliveira Neto

Additional information is available at the end of the chapter

http://dx.doi.org/10.5772/54094

1. Introduction

The water availability is considered the climatic factor with large effect on agricultural productivity, being responsible to determine species distribution in different climate zones around the globe [1]. Effects of drought depend of plant development stage, intensity, and duration of the water restriction. In other hand, plant adaptive strategies will determine the tolerance level, and consequently your survival on these conditions of inadequate water supply [2].

Water deficit is an abiotic factor that affects the agricultural production with high frequency and intensity, influencing aspects related to plant development, such as decrease in photosynthesis rate, reduction in leaf area [3], and stomata closing [4]. Crops normally present performance affected by water deficiency, which can cause lower growth and development (Figure 1), with progressive reduction in leaf dry matter [5] and consequent repercussion on production parameters, such as number of grains and pods per plant.

Root system presents complex strategy aiming to maintain water supply in conditions of water deficit, by increasing the root elongation rate and completely inhibiting the shoot [6]. On the other hand, plants growing in low water potentials normally present root thinner [7], and this morphological modification is an adaptation to increase water absorption efficiency. Therefore, a combination of changes in morphological, physiological and biochemical levels are necessary to plant survival in environments affected by drought.

Figure 1. Visual aspect of shoot in *Phaseolus vulgaris* plants exposed to drought by four days.

The biological fixation of nitrogen is the capacity of an organism to divide the molecule of nitrogen (N_2) and to combine hydrogen atoms (H^+), forming ammonium (NH_4^+) [8], being carried out by a distinct group of microorganisms, singly or under symbiosis. The *Bradyrhizobium* and *Rhizobium* genders are described as soil bacteria that have ability to infect root hair of leguminous plants, and it to induce nodule formation (Figure 2), with subsequent fixation of nitrogen [9].

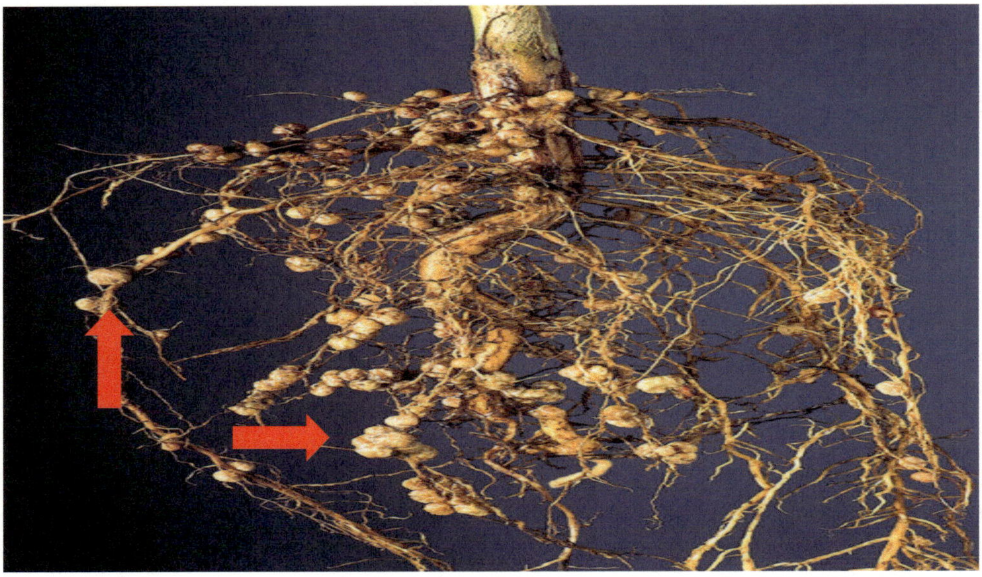

Figure 2. Visual aspect of root system of *Vigna unguiculata* plants inoculated with *Bradyrhizobium*. The red arrows indicate nodules formed after infection process.

http://dx.doi.org/10.5772/54094

The persistence of rhizobial strains, and their symbiotic performance in current and subsequent seasons are affected by numerous biotic and abiotic factors [10], with drought stress and nitrogen deprivation, being among the most significant in many parts of the world [11]. Other important factor is the root exudation ability, which it will determine plant microbe associations so that the survival and tolerance of rhizobia during water restriction.

Molybdenum is an essential element for soil microorganisms, since it serves as a cofactor for different enzymes involved in the metabolism of nitrogen, carbon and sulfur. Before the synthesis of molybdoenzymes, uptake of molybdate, which is the more stable form of molybdenum), its activation to an appropriate form, and its incorporation into the organic fraction of the molybdenum-cofactors, are required [12].

The presence of molybdenum is necessary during formation of several proteins, including the nitrogenase, the molybdoenzyme that reduces atmospheric dinitrogen (N_2) into ammonia (NH_4^+) [13]. This bacterium is also capable of denitrification, via nitric oxide (NO) and nitrous oxide (N_2O) to N_2, when the cells are cultured under oxygen-limiting conditions [14]. The first reaction of denitrification, is carried out by the periplasmic Mo-containing nitrate reductase [15]. In addition, the reaction under normal conditions is described as $N_2 + 8\ e^- + 8\ H^+ + 16\ MgATP \rightarrow 2\ NH_3 + H_2 + 16\ MgADP + 16Pi$.

The enzyme mechanism requires reduction of the Fe protein by electron donors such as ferredoxin and flavodoxin, transfer of single electrons from the Fe protein to the MoFe protein (which is dependent on MgATP hydrolysis) and, finally, internal electron transfer in the MoFe protein by the P cluster to the FeMo cofactor substrate-binding site. Each electron-transfer step requires an obligatory cycle of association of the Fe and MoFe proteins to form a complex (Figure 3), after which the two components dissociate [16].

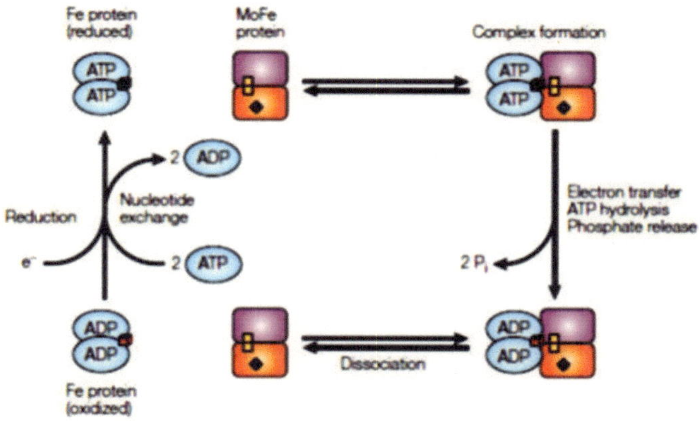

Figure 3. Schematic representation of the nitrogenase Fe protein cycle. The Fe protein dimer is shown in light blue with the cube representing the [4Fe–4S] cluster coloured black to indicate the reduced form and red to represent the oxidized form. The α and β subunits of the MoFe protein are depicted as orange and pink, respectively, the yellow squares represent the P cluster and the black diamond represents the FeMo cofactor. Changes in the oxidation state of the MoFe protein are not shown [16].

Several leguminous such as *Vigna unguiculata* and *Cicer arietinum* are considered tolerant to water deficit, and important mechanisms were developed by this species to tolerate inadequate water supply. For example, biochemical modifications in carbon metabolism, such as increase in sucrose [17], as well as significant interference in nitrogen metabolism, like reduction of soluble proteins [5] and increase in total amino acids [18] contribute to osmotic adjustment of these plants (Figure 4).

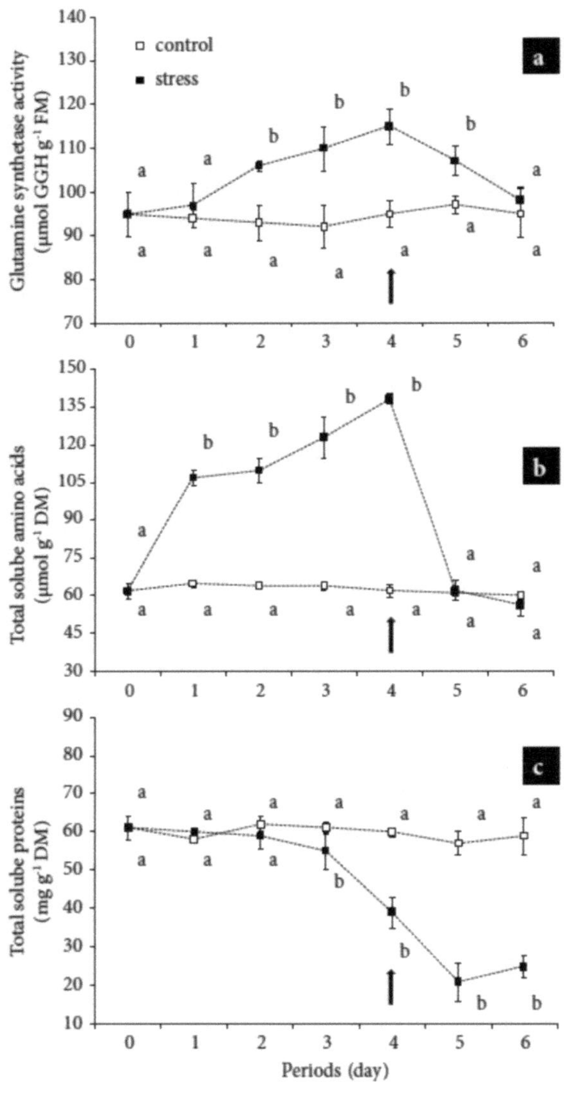

Figure 4. Glutamine synthetase activity (a), total soluble amino acids (b) and total soluble proteins (c) in *Vigna unguiculata* plants cv. Vita 7 subjected to 4 days of water restriction and 2 days of rehydration. Means followed by the same letter are not significantly different by the Tukey test at 5% of probability. The bars represent the mean standard error and the arrow the arrow indicates the rehydration point [5].

2. Objective

Aims of this chapter is to define (i) water deficiency and biological fixation of nitrogen, to explain (ii) as this symbiotic process can promote beneficial repercussions to plant and microorganism, and to present (iii) the attenuation of negative impacts on nodule and plant, besides nitrogen compounds and morphological parameters of plants exposed to water restriction.

3. Water maintenance in leaf and nodule produced by inoculation

Drought is environmental component that affect crop yields worldwide. In nature, this stress is multifaceted problems that are usually associated with other adverse circumstances, which limit plant performance such as water shortage and nutrient deficits. In order to assess the osmotic stress, Sassi et al. [19] monitored two *Phaseolus vulgaris* cultivars inoculated with *Rhizobium*, being cvs. 'Flamingo' (tolerant) and Cv. 'Coco Blanc' (sensitive).

Leaf osmotic potential (Ψo) decreased in stressed plants in both cultivars. A minimum value of −2.3 MPa was reached in Cv. 'Flamingo' plants under mannitol-induced osmotic stress (Figure 5 A). Ψo decreased in stressed nodules, reaching −1.3 MPa in Cv. 'Coco Blanc' and −1.7 MPa in Cv. 'Flamingo' (Figure 5 B). Therefore, Cv. 'Flamingo' showed a better osmotic adjustment response to osmotic stress both in leaves and nodules [19].

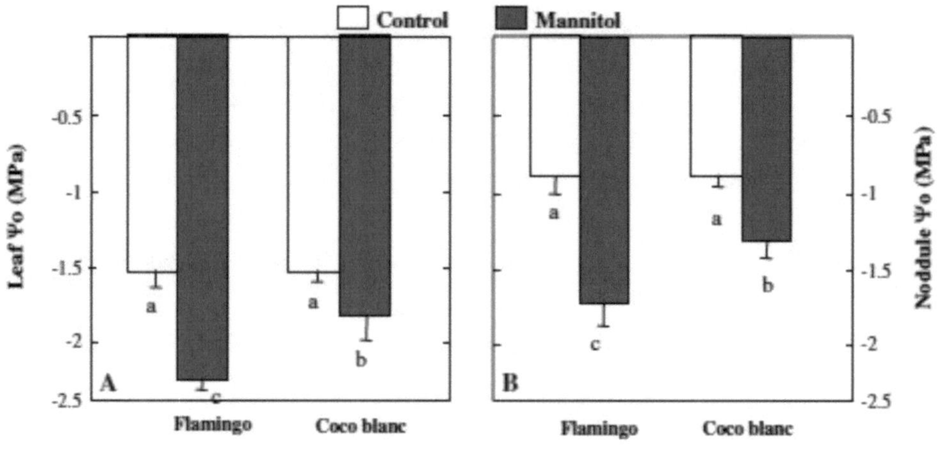

Figure 5. Variation of osmotic potential (Ψo) in response to osmotic stress in leaves (A) and nodules (B) mediated by 50 mM mannitol. Values represents mean ± SE (n=6) [19].

In control leaves of both cultivars, RWC remained close to 80% (Figure 6 A). After 15 days of osmotic treatment, RWC was 65%in mannitol-treated plants of Cv. 'Flamingo', and only

45% in Cv. 'Coco Blanc'. These results indicate that osmotic stress caused an important reduction in shoot water supply. The same trend was observed in nodules (Figure 6 B). Indeed, data showed decreased nodule RWC in both stressed cultivars. This decrease was higher in Cv. 'Coco Blanc' treated nodules [19].

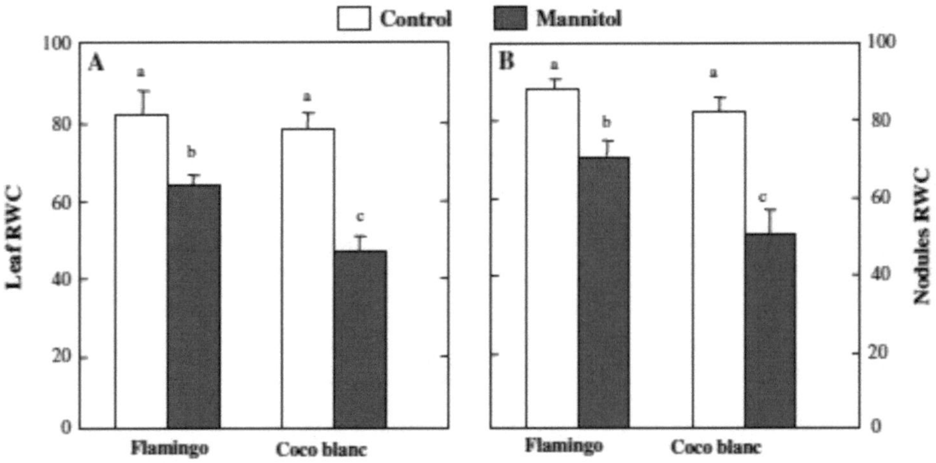

Figure 6. Effect of mannitol-induced osmotic stress on relative water content (RWC) in Flamingo and Coco blanc bean cultivar leaves (A) and nodules (B). Values represents mean ± SE (n=6) [19].

Mannitol-induced water deficit produced substantial dehydration that led to decreasing Ψo (Figure 6). The decrease in Ψo is considered a potential mechanism of cellular drought resistance as it enables turgor maintenance and growth continuation [20]. Cv. 'Flamingo' exhibited lower Ψo under osmotic treatment. It was able to uptake more water and then grow more when exposed to decreased Ψo, thus it turned out to be a better drought tolerant cultivar than Cv. 'Coco Blanc' [21]. This may be attributed to maintenance of the leaf and nodule water status under stressed conditions (Figure 5). Several mechanisms could be involved in contributing to water retention.

4. *Bradyrhizobium* ameliorates negative effects in plants exposed to drought

The relationship between the water status in the plant and N2 fixation, mainly under water stress, and the changes in nodule morphology have been studied in some temperate legumes [22]. However, tropical legumes growing in arid regions, have not received adequate attention.

Even where information is available, the degree of water stress in the plants was not clearly defined, which makes it difficult to make comparisons. The structural basis for the difference in sensivity of N2 fixation in tropical legumes, under water stress, is not clearly understood [23]. Based in these problems reported, Figueiredo et al. [24] investigated Vigna unguiculata plants exposed to 3 inoculation forms (BR-2001, EI-6, and control) combined with 6 different degrees of water stress (-1.5, -2.0, -4.0, -6.0, -8.0, and -10.0 kPa).

Water deficit response in cowpea appears to be directly related to a reduction in nodule mass (Table 1), which may (after a severe stress, S6) have affected nodule structural constituents. However, in moderate stress (S3) the impact on nodule water content was higher than on the changes in nodule mass [24].

Matric potentials[(1)] (kPa)	LHb ($mg\ g^{-1}$ nod.DM)			UN ($nmol\ ml^{-1}$)			N_2ase ($nmol\ C_2H_4\ pl^{-1}\ h^{-1}$)		
	Strains								
	BR-2001	EI-6	C	BR-2001	EI-6	C	BR-2001	EI-6	C
S1	26.16ab^A	28.06ab^A	23.00ab^B	5179a^A	5351ab^A	4000a^B	10697a^A	11274b^A	9288a^B
S2	28.56a^A	29.70a^A	24.23a^B	5840a^A	6099a^A	4402a^B	11670a^A	12410a^A	9571a^B
S3	25.40b^A	27.23ab^A	20.40bc^B	3740b^B	4488bc^A	2906b^C	2563b^B	3400c^A	2093b^B
S4	25.40b^A	26.50b^A	20.40bc^B	3366bc^B	4114cd^A	2733b^B	2403b^{AB}	2633cd^A	1630bc^B
S5	24.03b^A	25.90b^A	20.30bc^B	2819bc^B	3567cd^A	2388b^B	1130c^B	2277d^A	1000c^B
S6	23.36b^B	25.90b^A	19.26c^C	2589c^B	3337d^A	2071ab^B	1121c^{AB}	1756d^A	837c^B
F (plot)		13.90**			132.87**			142.15**	
F (subplot)		132.36**			67.67**			67.48**	
% CV (plot)		5.54			7.44			23.94	
% CV (subplot)		4.57			9.54			20.25	

(1)For S1 to S6 see Table 1. *, **Significant at the 0.05 and 0.01 probability level. In each column (lower letters) and in each line (capital letters), the means followed by the same letter do not differ statistically ($p < 0.05$) from each other, according to Tukey's test

Table 1. Nodule dry matter (NDM) and nodule water content (NWC) in cowpea with (BR-2001 and EI-6) and without (C) *Bradyrhizobium* spp. inoculation at different degrees of water stress [24].

5. Interference positive on nitrogen compounds of plants inoculated and exposed to water deficit

Cowpea (*Vigna unguiculata* [L.] Walp.) is a leguminous with high protein content, large capacity of fixation of the atmospheric nitrogen (N_2) and low requirements to soil fertility [25], being frequently cultivated by farmers in Northern and Northeastern regions of the Brazil. This species constitutes the main subsistence culture, being the grain used as protein source in feeding [26]. Cowpea presents important agronomical characteristics, such as rusticity and precocity, besides being considered a plant adapted to conditions of limited or insufficient water availability [27].

Beneficial effects proportioned by the inoculation on growth parameters as leaf, stem and root are largely explored and well known in leguminous plants [28-30], but informations more specific of this symbiotic process on essential compounds such as amino acids and proteins are limited. Figueiredo et al. [24] report that inoculation using *Bradyrhizobium* can alleviate the negative consequences in *Vigna unguiculata* plants induced to water deficiency, but study conducted by Serraj and Sinclair [31] revealed that water supply presents repercussion on symbiotic efficiency.

Based on this overview, Barbosa et al. [32] carried out a study aiming to investigate if nitrogen compounds exercise influence on accumulation of dry matter in *Vigna unguiculata* plants exposed to combined action of inoculation and water deficit.

The concentration of total soluble amino acids in plants subjected to inoculation was higher only in tolerant plants, if compared with same treatments of plants non-inoculated (Figure 7 A). Water deficit promoted a significant increase in this variable to all treatments. The tolerant cultivar presented lower changes, in comparison with same treatments in sensitive cultivar.

Total soluble proteins of inoculated plants presented higher values (Figure 7 B), when compared to same treatments in non-inoculated plants. Water deficit caused a significant decrease in both cultivars, presenting higher variation in sensitive plants.

For proline the inoculated plants presented higher values, comparing with same treatments in non-inoculated plants (Figure 7 C). The two cultivars demonstrated higher values in water deficit, when compared with respective controls. These results present a greater variation in tolerant plants, if compared with same treatments in sensitive plants.

Tolerant plants submitted to inoculation presented significant increase in amino acids, and these results are attributed to biological fixation of nitrogen. The nitrogenase enzyme promotes the nitrogen absorption in form of nitrogen gas (N_2) and conversion to ammonium (NH_4^+). In addition, the higher formation of amino acids probably is linked to increase in activity of enzymes glutamine synthetase (GS), being your activity depending of ATP (adenosine-5'-triphosphate), and glutamate synthase (GOGAT). In addition, the increase in amino acids of plants exposed to inoculation is due to greater flux and better assimilation of nitrogen in form of ammonium, concomitantly with higher activity of GS and GOGAT enzymes. Ramos et al. [33] evaluating the responses in Glycine max plants under water deficit and inoculation of Bradyrhizobium japonicum observed also an increase in concentration of total soluble amino acids.

The concentration of total soluble amino acids in plants under water deficit increased in all treatments. This increment occurred probably due to increase in activity of protease enzymes, responsible by breakdown of proteins aiming to adjust osmotically the plant [34]. Similar results on increase in amino acids were obtained to Costa et al. [35] investigating *Vigna unguiculata* plants. Delfini et al. [36] evaluating the responses of two *Arachis hypogaea* cultivars submitted to inoculation of *Bradyrhizobium sp.* showed significant increase in amino acids.

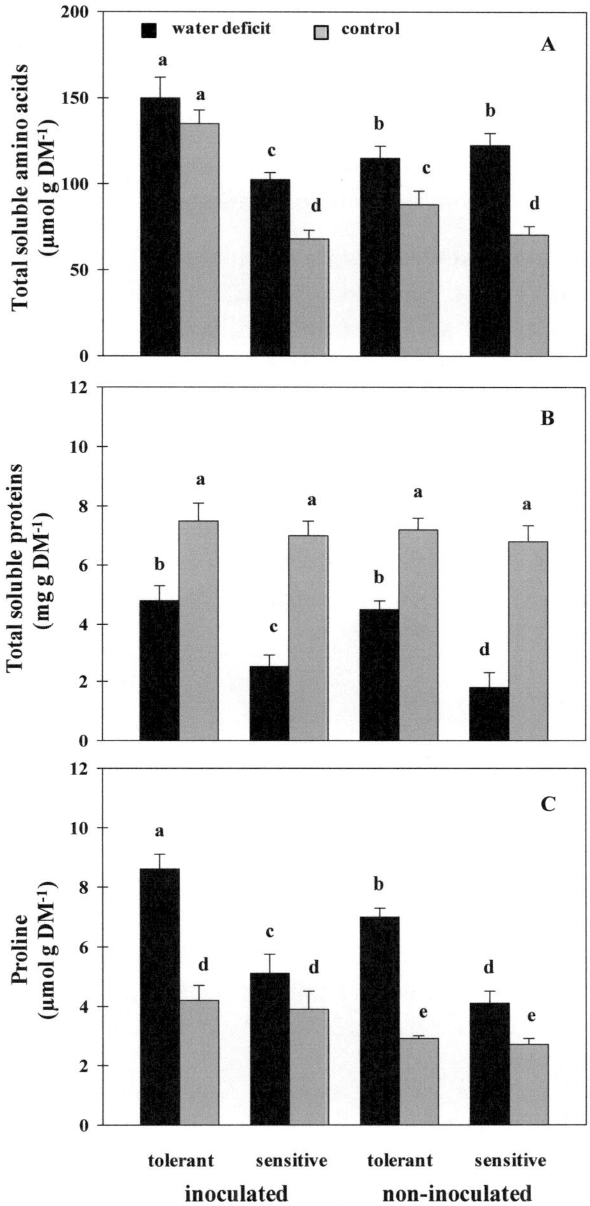

Figure 7. Total soluble amino acids (A), total soluble proteins (B), and proline (C) in two contrasting *Vigna unguiculata* plants under water deficit and subjected to inoculation. Means followed by the same letter are not significantly different by the Scott-Knott test at 5% of probability. The bars represent the mean standard error [32].

The increase showed in total soluble proteins induced by inoculation suggests that bacteria action resulted in increase in nitrogen supply through secondary route, that is regulated by the nitrogenase [37], because in this study was not verified increase in activity of the nitrate reductase after inoculation. Hristozkova et al. [38] evaluating the responses in *Pisum sativum* plants under inoculation and molybdenum application also obtained increase in protein levels.

The decrease in protein levels promoted by the water deficit is associated to decrease of the protein synthesis combined with increase of proteolytic enzymes, responsible by breakdown of soluble proteins in plants [39]. Costa [40] obtained similar results studying *Vigna unguiculata* subjected to water deficit, corroborating these results.

The increase of proline levels provoked by the inoculation is probably linked to better amino acids utilization such as glutamic acid and arginine, being the glutamic acid the precursor of the proline, while arginine can suffer reaction mediated by enzyme called of pyrrolline-5-carboxylate reductase (P5CR) and consequently to liberate proline [33].

Kohl et al. [41] also observed higher amounts of proline in *Glycine max* plants inoculated with *Bradyrhizobium japonicum*, contributing with results of this study. The increase of proline in plants under water deficit is a response to loss of cell turgescence [42]. Nogueira et al. [43] describe that the proline accumulation has been related with drought tolerance in higher plants, actuating as osmoregulator agent with the objective to keep water in plant tissue [44]. Similar behavior was described by González et al. [45] working with *Pisum sativum* plants under water restriction.

6. *Bradyrhizobium* producing better performance on morphological parameters

Beneficial effects proportioned by the inoculation on growth parameters are largely explored in crops as *Phaseolus vulgaris* and *Glycine max*, but informations on dry matter accumulation of *Vigna unguiculata* under water deficit is limited. Barbosa et al. [32] conduced an experiment with 2 cultivars (tolerant and sensitive) combined with 2 water regimes (water deficit and control), and 2 inoculation forms (inoculated and non-inoculated), totalizing 8 treatments.

In shoot dry matter the inoculation provoked increase, considering same treatments in plants non-inoculated (Figure 8 A). However, this increase only was significant in tolerant cultivar under inoculation and irrigation. Water deficit occasioned a significant decrease in shoot dry matter, with exception in tolerant cultivar non-inoculated. The tolerant plants presented better results, if compared with sensitive plants independently of treatment (Figure 8 A).

The inoculation provoked increase in plant dry matter, with exception in sensitive cultivar under irrigation (Figure 8 B), comparing to same treatments in plants non-inoculated. Water deficiency induced decrease in plant dry matter for all treatments, being these significant results when compared with control plants. Independently of treatments was showed that sensitive cultivars presented lower values, if compared to tolerant cultivars.

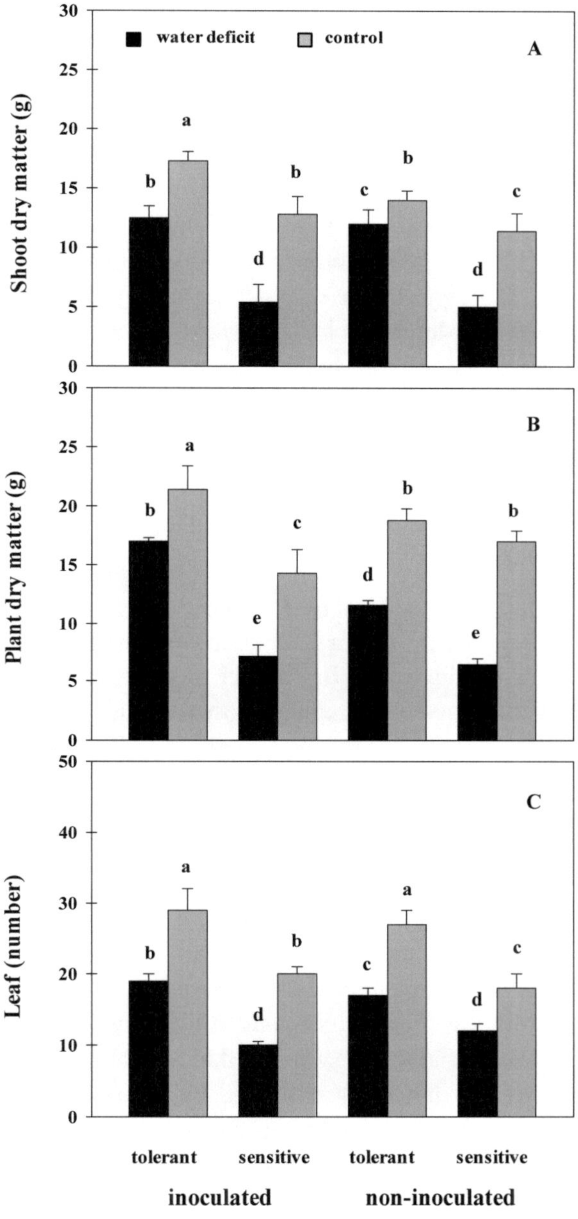

Figure 8. Shoot dry matter (A), plant dry matter (B), number of leaves (C) in two contrasting Vigna unguiculata plants under water deficit and subjected to inoculation. Means followed by the same letter are not significantly different by the Scott-Knott test at 5% of probability. The bars represent the mean standard error [32].

For leaf number occurred increase after inoculation, with exception in sensitive plants under water deficiency (Figure 8 C). Water deficit proportioned decrease in values of leaf number, being significant in comparison with control plants. In tolerant cultivar were obtained higher values of leaf number, if compared with sensitive cultivar.

Shoot dry matter was maximized after the inoculation procedure, being this fact linked to probably increase in nodule number in root (data not shown), as well as it proportioned higher absorption and availability of nitrogen to plant [46-47]. Similar results linked to shoot dry matter were found by Figueiredo et al. [24] in research with *Vigna unguiculata* plants exposed to inoculation of *Bradyrhizobium*.

The water deficit reduced the production of shoot dry matter, with these effects associated to negative interference of water deficiency on biochemical processes as nitrate assimilation and biological fixation of nitrogen [35], modifying indirectly the partitioning of photo-assimilates in root and shoot, and consequent decrease in accumulation of shoot biomass [48]. Similar results were found by Mendes et al. [49] working with two *Vigna unguiculata* cultivars submitted to drought during two stages.

The inoculation proportioned increase of total dry matter, and this result must be linked to better development and efficiency of root system, in which presents higher nitrogen absorption using the nodulation process. I addition, normally the higher nitrogen fixation will produce increase in amino acids and also proteins [36], and it exercises influence on photoassimilates availability.

Similar responses were described by Sassi et al. [50] investigating two *Phaseolus vulgaris* cultivars subjected to inoculation with bacteria of *Rhizobium* gender. Plants under water deficiency frequently have the production of dry matter reduced, being this decrease related to fact that water deficit affects several metabolic processes such as absorption of water and nutrients, which are fundamental to keep adequate growth and development rates.

Nascimento [51] also reported that water deprivation affects the osmotic mechanism, and by consequence reduces the CO_2 supply, that is essential in photosynthetic process. Similar results were found by Leite and Virgens Filho [52] studying *Vigna unguiculada* plants exposed to water deficit.

The increase in leaf number promoted by inoculation is occasioned by the higher number of nodules in root, and consequently due to the better biological fixation of nitrogen [53]. Araújo et al. [54] studying *Vigna unguiculata* and *Leucaena leucocephala* plants also reported an increase of this variable, confirming the results obtained in this work. The lower leaf number after water deficiency is caused by the process of leaf abscission, and this fact occurs due to substrate not to present water and nutrient sufficient to supply the plant exigencies [4]. Correia and Nogueira [55] obtained similar results with *Arachis hypogaea* plants under water deficit.

7. Final considerations

This chapter was structured with recent informations on capacity of *Bradyrhizobium* and *Rhizobium* to mediate tolerance in leguminous plants submitted to water deficit, which it can be used by students, teachers, researchers, scientists and farmers. It revealed concepts, effects, and results on water deficiency and your consequences on plants, as well as explored several possibilities linked to symbiosis between plant-microorganisms. Additionally, it presented essential compounds such as molybdenum and reactions during process of biological fixation of nitrogen. Also was demonstrated the water maintenance in leaf and nodule produced after inoculation. Based in novel results, were related interference positives on nitrogen compounds such as total soluble amino acids, proline, and total soluble proteins. Other results prove the beneficial repercussion produced by inoculation with *Bradyrhizobium* on morphological parameters.

Acknowledgements

This chapter had financial support from Conselho Nacional de Pesquisa (CNPq/Brazil) and Coordenação de Aperfeiçoamento de Pessoal de Nível Superior (CAPES/Brazil) for Lobato AKS.

Author details

Allan Klynger da Silva Lobato[1], Joaquim Albenísio Gomes da Silveira[2],
Roberto Cezar Lobo da Costa[1] and Cândido Ferreira de Oliveira Neto[1]

1 Núcleo de Pesquisa Vegetal Básica e Aplicada, Universidade Federal Rural da Amazônia, Paragominas, Brazil

2 Laboratório de Metabolismo e Estresse de Plantas, Universidade Federal do Ceará, Fortaleza, Brazil

References

[1] Rockström J, Falkenmark M. Semiarid crop production from a hydrological perspective: Gap between potential and actual yield. Crit. Rev. Plant Sci 2000;19 319-326.

[2] Kramer PJ, Boyer JS. Water relations of plant and soils. Academic Press, New York. 1995.

[3] Fontana DC, Berlato MA, Bergamaschi H. Micrometeorological alterations in soybeans grown under different water regimes. Pesquisa Agropecuária Brasileira 1992;27 661-669.

[4] Santos RF, Carlesso R. Water deficit and morphologic and physiologic behavior of plants. Revista Brasileira de Engenharia Agrícola e Ambiental 1998;2 287-294.

[5] Costa RCL, Lobato AKS, Silveira JAG, Laughinghouse HD. ABA-mediated proline synthesis in cowpea leaves exposed to water deficiency and rehydration. Turkish Journal of Agriculture and Forestry 2011;35 309-317.

[6] Sharp RE, Poroyko V, Hejlek LG, Spollen WG, Springer GK, Bohnert HJ, Nguyen HT. Root growth maintenance during water deficits: Physiology to functional genomics. Journal of Experimental Botany 2004;55 2343-2351.

[7] Sharp RE, Silk WK, Hsiao TC. Growth of the maize primary root at low water potentials. I. Spatial distribution of expansive growth. Plant Physiology 1988;87 50-57.

[8] Albino UB, Campo RJ. Effect of sources and levels of molybdenum on *Bradyrhizobium* survival and on biological nitrogen fixation in soybean. Pesquisa Agropecuária Brasileira 2001;36 527-534.

[9] Mercante MM, Gol SR, Franco AA. Importance of phenolics compounds in the interactions between leguminous species and rhizobia. Ciências da Vida 2002; 22 65-81.

[10] Bordeleau LM, Prevost D. Nodulation and nitrogen fixation in extreme environments. Plant Soil 1994;116 115–125.

[11] Squartini A, Dazzo FB, Casella S, Nuti MP. The root-nodule symbiosis between "*Rhizobium hedysari*" and its drought-tolerant host *Hedysarum coronarium*. Symbiosis 1993;15 227–238

[12] Pau RN, Lawson DM. Transport, homeostasis, regulation, and binding of molybdate and tungstate to proteins. Metal Ions in Biological Systems 2002;39 31–74.

[13] Lawson DM, Smith BE. Molybdenum nitrogenases: a crystallographic and mechanistic view. Metal Ions in Biological Systems 2002;39 75–119.

[14] Bedmar EJ, Robles EF, Delgado MJ. The complete dentrification pathway of symbiotic N-fixing bacteria *Bradyrhizobium japonicum*. Biochemical Society Transactions 2005;1 11–16.

[15] Delgado MJ, Bonnard N, Tresierra-Ayala A, Bedmar EJ, Müller P. The *Bradyrhizobium japonicum napEDABC* genes encoding the periplasmic nitrate reductase are essential for nitrate respiration. Microbiology 2003;149 3395–3403.

[16] Dixon R, Kahn D. Genetic regulation of biological nitrogen fixation. Nature Reviews Microbiology 2004;2 621-631.

[17] Lobato AKS, Costa RCL, Oliveira Neto CF, Santos Filho BG, Alves GAR, Freitas JMN, Cruz FJR, Marochio CA, Coimbra GK. Responses of the pigments and carbon metabolism in *Vigna unguiculata* cultivars submitted to water deficit. Research Journal of Biological Sciences 2009a;4 593-598.

[18] Lobato AKS, Oliveira Neto CF, Costa RCL, Santos Filho BG, Cruz FJR, Laughinghouse IV HD. Biochemical and physiological behaviour of *Vigna unguiculata* (L.) Walp. under water stress during the vegetative phase. Asian Journal of Plant Science 2008a;7 44-49.

[19] Sassi S, Aydi S, Gonzalez EM, Arrese-Tgor C, Abdelly C. Understanding osmotic stress tolerance in leaves and nodules of two *Phaseolus vulgaris* cultivars with contrasting drought tolerance. Symbiosis 2010;52 1-10.

[20] Bajji M, Lutts S, Kinet JM. Physiological changes after exposure to and recovery from polyethylene glycol-induced water deficit in callus culture issued from durum wheat (*Triticum durum*) cultivars differing in drought resistance. Journal of Plant Physiology 2000;156 75–83.

[21] Sassi AS, Aydi S, Gonzalez EM, Abdelly C. Osmotic stress affects water relations, growth, and nitrogen fixation in *Phaseolus vulgaris* plants. Acta Physiologiae Plantarum 2008a;30 441–449.

[22] Sprent JJ. Nitrogen fixation. In Physiology and Biochemistry of drought resistance in plants. Eds. L C Paleg and D Aspinall. pp 131–143. Academic Press. 1981.

[23] Venkateswarlu B, Neelam S, Maheswari M. 1990 Nodulation and N_2 (C_2H_2) fixation in cowpea and groundnut during water stress and recovery. Field Crops Research 1990;25 223–232.

[24] Figueiredo MVB, Vilar JJ, Burity HA, França FP. Alleviation of water stress effects in cowpea by *Bradyrhizobium ssp*. Inoculation. Plant and Soil 1999;207 67-75.

[25] Alves JMA, Araújo NP, Uchôa SCP, Albuquerque JAA, Silva AJ, Rodrigues GS, Silva DCO. The agroeconomic evaluation of the production of cowpea varieties intercropped with varieties of cassava in The State of Roraima. Revista Agroambiente 2009;3 15-30.

[26] Frota KMG, Soares RAM, Arêas JAG. Chemical composition of cowpea [*Vigna unguiculata* (*L.*) *Walp*], BRS-Milênio cultivar. Ciência e Tecnologia de Alimentos 2008;28 470-476.

[27] Oliveira AP, Sobrinho JT, Nascimento JT, Alves AU, Albuquerque IC, Bruno GB. Evaluation of breeding lines and cultivars of cowpea-beans in Areia, Paraiba, Brazil. Horticultura Brasileira 2002;20 180-182.

[28] Ramos MLG, Gordon AJ, Minchin FR, Sprent JI, Parsons R. Effect of water stress on nodule physiology and biochemistry of a drought tolerant cultivar of common bean (*Phaseolus vulgaris L.*). Annals of Botany 1999;83 57-63.

[29] Silveira JAG, Matos JCS, Cecatto VM, Viegas RA, Oliveira JTA. Nitrate reductase activity, distribution, and response to nitrate contrasting *Phaseolus* species inoculated with *Rhizobium* spp. Environment and Experimental Botany 2001;46 37- 46.

[30] Moraes WB, Martins Filho S, Garcia SGO, Caetano SP, Moraes WB, Cosmi FC. Evaluation of biological, fixation of nitrogen in *rhizobium* under water deficit. Idesia 2010;28 61-68.

[31] Serraj R, Sinclair TR. Processes contributing to N_2 fixation insensitivity to drought in the soybean cultivar Jackson. Crop Science 1996;36 961-968.

[32] Barbosa MAM, Lobato AKS, Viana GDM, Coelho KNN, Barbosa JRS, Moraes MCHS, Costa RCL, Santos Filho BG, Oliveira Neto CF. Relationship between total soluble proteins and dry matter in two contrasting cowpea cultivars induced to inoculation and water deficiency. The Scientific World Journal, 2012;paper is press.

[33] Ramos MLG, Parsons R, Sprent JI. Differences in ureide and amino acid content of water stressed soybean inoculated with *Bradyrhizobium japonicum* and *B. elkanii*. Pesquisa Agropecuária Brasileira 2005;40 453-458.

[34] Costa RCL, Lobato AKS, Oliveira Neto CF. Variation in content of total soluble amino acids in leaves of cowpea under water stress. In: Congresso Nacional de feijão caupi, pp. 1-19, Teresina, PI, maio, 2006a.

[35] Costa RCL, Cardoso BB, Silva JT, Gomes Filho JGF, J. A. G. Silveira JAG. Water stress strongly decreases the assimilation of nitrate and nodulation in cowpea (*Vigna unguiculata,* (L.) Walp.). *In* Reunião Nacional de Pesquisa de Caupi, pp. 78-79, Teresina, PI, Brazil, march, 1996.

[36] Delfini R, Belgoff C, Fernández E, Fabra A, Castro S. Symbiotic nitrogen fixation and nitrate reduction in two peanut cultivars with different growth habit and branching pattern structures. Plant Growth Regulation 2010;6 153-159.

[37] Marschner H, Mineral nutrition of higher plants, Academic Press, pp. 889, London, UK, 1995.

[38] Hristozkova M, Geneva M, Stancheva I, Response of inoculated pea plants (*Pisum sativum* L.) to root foliar fertilizer application with reduced molybdenum concentration. Plant Physiology 2006;45 73-79.

[39] Lechinoski A, Freitas JMN, Castro DS, Lobato AKS, Oliveira Neto CF, Cunha RLM. Influence of water stress on levels of soluble amino acids and proteins in leaves of teak (*Tectona grandis L. f.*). Revista Brasileira de Biociência 2007;5 927-929.

[40] Costa RCL. Nitrogen assimilation and osmotic adjustment in plants nodulated bean-to-string *Vigna unguiculata* (L.) Walp. Subjected to water stress. Ph.D Thesis. Universidade Federal do Ceará, Brasil, 1999.

[41] Kohl DH, Kennelly DJ, Zhy Y, Schubert KR, Sheader G. Proline accumulation, nitrogenase (C_2H_2 reducing) activity and activities of enzymes related to proline metabolism in drought-stressed soybean nodules. Journal of Experimental Botany 1991;42 831-837.

[42] Oliveira AAO, Barreto LP, Bezerra Neto E, Santos MVF, Costa JCA. Organic solutes in forage sorghum genotypes under salt stress. Pesquisa Agropecuária Brasileira 2006;14 31-35.

[43] Nogueira RJMC, Santos CR, Neto EB, Santos VF. Physiological behaviour of two peanut cultivars submitted to suppression watering. Pesquisa Agropecuária Brasileira 1998;33 1963-1969.

[44] Muchow RC, Carberry PS. Designing improved plant types for the semi-arids tropics: Agronomist` viewpoints. In: penning de vries, F.W.T. Teng, P. Metselaar, Eds., Dordrecht:Kluwer, pp. 37-61, Meryland, EUA, 1993.

[45] González EM, Aparicio-Tejo PM, Gordon AJ, Minchin FR, Igor CA. Water-deficit effects on carbon and nitrogen metabolism of pea nodules. Journal of experimental Botany 1998;49 1705-1714.

[46] Epstein E, Bloom AJ. Nutrition and growth. In: *E.* Epstein and A. J. Bloom, Ed. pp. 251-286, Londrina, PR, Brazil, 2006.

[47] Souza RA, Hungria M, Franchini JC, Maciel CD, Campo RJ, Zaia DAM. Minimal set of parameters for evaluation soil microbiota and biological nitrogen fixation in soybean. Pesquisa Agropecuária Brasileira 2008;43 83-91.

[48] Correia GK, Nogueira CMJR. Evaluation of the growth of groundnut (*Arachis hypogae L.*) subjected to water deficit. Revista de Biologia e Ciências da Terra 2004a;4 56-60

[49] Mendes RMS, Távora FJAF, Pitombeira JB, Nogueira RJMC. Source-sink relationships in cowpea under drought stress. Revista Ciência Agronômica 2007;38 95-103.

[50] Sassi S, Aydi S, Gonzalez EM, Arrese-Tgor C, Abdelly C. Understanding osmotic stress tolerance in leaves and nodules of two *Phaseolus vulgaris* cultivars with contrasting drought tolerance. Symbiosis 2010;52 1-10.

[51] Nascimento SP. Effect of water deficit in cowpea to identify genotypes with drought tolerance. Ph.D Thesis, Universidade Federal do Piauí, 2009.

[52] Leite ML, Virgens Filho JS. Dry matter production of cowpea [*Vigna unguiculata* (L.) Walp] plants submitted to water deficits. Exatas Terra 2004;10 43-51.

[53] Ferreira EPB, Martins LMV, Xavier GR, Rumjanek NG. Nodulation and grain yield by cowpea [*Vigna unguiculata* (L.) Walp.] Inoculated with rhizobia, isolates. Revista Caatinga 2011;24 27-35.

[54] Araújo ASF, Carneiro RFV, Bezerra AAC, Araújo FF. Co-inoculation *Rhizobia* and *Bacilus subtilis* in cowpea and Leucaena: effects on nodulation, N_2 fixation and plant growth. Ciência Rural 2009;40 43-48.

[55] Correia KG, Nogueira RJMO. Assessment of growth of peanut (*Arachis hypogaea L.*) subjected water deficit. Revista de Biologia e Ciência da Terra 2004b;4 240-248.

Chapter 4

Comparison Between the Water and Salt Stress Effects on Plant Growth and Development

Alexandre Bosco de Oliveira,
Nara Lídia Mendes Alencar and Enéas Gomes-Filho

Additional information is available at the end of the chapter

http://dx.doi.org/10.5772/54223

1. Introduction

Abiotic stress limits crop productivity [1], and plays a major role in determining the distribution of plant species across different types of environments. Abiotic stress and its effects on plants in both natural and agricultural settings is a topic that is receiving increasing attention because of the potential impacts of climate change on rainfall patterns and temperature extremes, salinization of agricultural lands by irrigation, and the overall need to maintain or increase agricultural productivity on marginal lands. In the field, a plant may experience several distinct abiotic stresses either concurrently or at different times through the growing season [2].

In reference [3] are showed some common examples of the abiotic stresses a plant may encounter which include a decreased availability of water, extremes of temperature including freezing, decreased availability of essential nutrients from the soil (or conversely the build-up of toxic ions during salt stress), excess light (especially when photosynthesis is restricted) or increased hardness of the soil that restricts root growth.

In the face of a global scarcity of water resources and the increased salinization of soil and water, abiotic stress is already a major limiting factor in plant growth and will soon become even more severe as desertification covers more and more of the world's terrestrial area. Plants are often subjected to periods of soil and atmospheric water deficits during their life cycle. Moreover, the faster-than-predicted change in global climate [4] and the different available scenarios for climate change suggest an increase in aridity for the semiarid regions of the globe. Together with overpopulation, this will lead to an overexploitation of water resources for agriculture purposes and increased constraints on plant growth and survival aid, therefore, on realizing crop yield potential [5]. Thus, if a single abiotic stress is to be identified as the most common in limiting the growth of crops worldwide, it most probably is low water supply [1].

Like the water stress, salinity is one of the major severe abiotic factors affecting crop growth and productivity [6]. Salt's negative effects on plant growth have initially been associated with the osmotic stress component caused by decreases in soil water potential and, consequently, restriction of water uptake by roots.

The literature shows that drought and salinity are already widespread in many regions. Therefore, in reference [7] the authors have presented that in world where population growth exceeds food supply, agricultural and plant biotechnologies aimed at overcoming severe environmental stresses need to be fully implemented.

1.1. Plant stress definitions

The term stress is most often used subjectively and with various meanings. The physiological definition and appropriate term for stress are referenced as responses to different situations. The flexibility of normal metabolism allows the development of responses to environmental changes, which fluctuate regularly and predictably over daily and seasonal cycles [8]. Thus, stress plays a pivotal role in determining interaction outcomes because it strongly influences the strength of underlying positive and negative interactions.

Stress is defined as any external abiotic (heat, water, salinity) or biotic (herbivore) constraint that limits the rate of photosynthesis and reduces a plant's ability to convert energy to biomass [9]. The strength of positive interactions increases with increasing stress except at the most extreme levels. In contrast, the strength of negative interactions is either unrelated to stress and remains consistently high, or alternatively, decreases with increasing stress [10].

Environmental stress could be defined in plants as any change in growth condition(s), within the plant's natural habitat, that alters or disrupts its metabolic homeostasis. Such change(s) in growth condition requires an adjustment of metabolic pathways, aimed at achieving a new state of homeostasis, in a process that is usually referred to as acclimation [11]. However, the concept of plant stress is often used imprecisely, and stress terminology can be confusing, so it is useful to start our discussion with some definitions. Stress is usually defined as an external factor that exerts a disadvantageous influence on the plant. In most cases, stress is a measured in relation to plant survival, crop yield, growth (biomass accumulation), or the primary assimilation processes (CO_2 and mineral uptake), which are related to overall growth. In addition, the concept of stress is intimately associated with that of stress tolerance, which is the plant's fitness to cope with an unfavourable environment. In the literature the term stress resistance is often used interchangeably with stress tolerance, although the latter term is preferred [12].

According with the literature [13], changes in strength of these underlying processes drive shifts between competition and facilitation. The stress gradient hypothesis (SGH) predicts that facilitation and competition vary inversely along stress gradients with facilitation more frequent and stronger when stress is high and competition more frequent and stronger when stress is low. The SGH also predicts that the strongest facilitation should occur with competitive species at the upper limits of their stress tolerance while the strongest competition should occur with stress tolerant species located at their ecological optimum.

Shifts in the structure of interaction outcomes, i.e. a shift from competition to facilitation, along stress gradients are likely to have profound implications for community stability [14, 15]. Community compositional instability can be defined as changes in species abundances that drive directional changes in community composition. There is growing evidence that changes in the structure of species interactions can reduce such stability [16].

In both natural and agricultural conditions, plants are frequently exposed to environmental stresses. In [17] the work presents how some environmental factors, such as air temperature, can become stressful in just a few minutes; others, such as soil water content, may take days to weeks, and factors such as soil mineral deficiencies can take months to become stressful.

1.2. Concepts and consequences of water and salt stress on plants

Water-deficit stress can be defined as a situation in which plant water potential and turgor are reduced enough to interface with normal functions. Water stress is considered to be a moderate loss of water, which leads to stomatal closure and limitation of gas exchange. Desiccation is a much more extensive loss of water that can potentially lead to gross disruption of metabolism and cell structure and eventually to the cessation of enzyme catalyzing reactions. Water stress is characterized by reduction of water content, turgor, total water potential, wilting, closure of stomata, and decrease in cell enlargement and growth. Severe water stress may result in arrest of photosynthesis, disturbance of metabolism, and finally death [8, 18].

The term 'salinity' refers to the presence in soil and water of electrolytic mineral solutes in concentrations that are harmful to many agricultural crops. Except along seashores, saline soils seldom occur in humid regions, thanks to the net downward percolation of fresh water though the soil profile, brought about by the excess of rainfall compared with evapotranspiration. In arid regions, on the other hand, there may be periods of no net downward percolation and hence no effective leaching, so salts can accumulate in the soil. Hence the combined effect of meager rainfall, high evaporation, the presence of salt-bearing sediments, and – in many places, particularly river valleys and other low-lying areas – the occurrence of shallow, brackish groundwater which gives rise to saline soils [19].

Salinity in soil or water is one of the major stresses and, especially in arid and semi-arid regions, can severely limit crop production. The Figure 1 is showing that deleterious effects of salinity on plant growth are associated with low osmotic potential of soil solution (water stress), nutritional imbalance, specific ion effect (salt stress), or a combination of these factors [20].

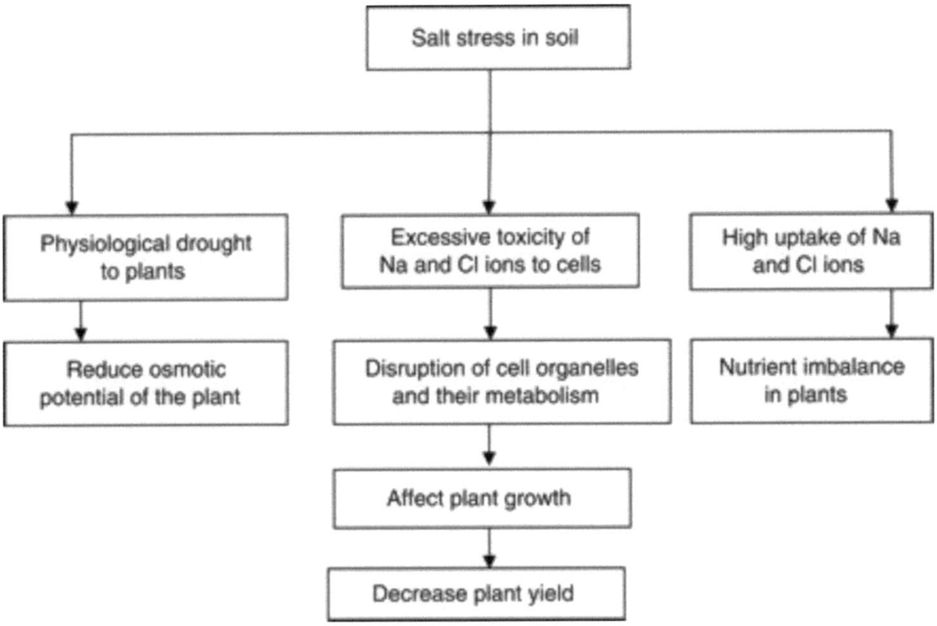

Figure 1. Effects of salt stress on plants (reproduced from [21]).

1.3. Mechanisms of acclimation or adaptation to water and salt stress

Drought and soil salinity are among the most damaging abiotic stresses affecting today's agriculture. It is understandable that plants are under periodic water stress because of the unpredictable nature of rainfall. Salt stress may also occur in areas where soils are naturally high in salts and/or where irrigation, hydraulic lifting of salty underground water, or invasion of seawater in coastal areas brings salt to the surface soil that plants inhabit. Plants have evolved mechanisms that allow them to perceive the incoming stresses and rapidly regulate their physiology and metabolism to cope with them. Very often such regulations and responses include feed-forward mechanisms for stress reduction that are in addition to the responses that are seen after stresses have caused irreversible damage to physiological functions. A good example of such a feed-forward mechanism the ability of plants to regulate their water loss through partial closure of stomata and/or reduced leaf development, long before there is a substantial loss of their leaf turgor or some irreversible damage to inner membrane systems. [22-24].

In this way, talking specific about water deficit, the physiological responses of plants to water stress include leaf wilting, a reduction in leaf area, leaf abscission, and the stimulation of root growth by directing nutrients to the underground parts of the plants. Plants are more susceptible to drought during flowering and seed development (the reproductive stages), as plant's resources are deviated to support root growth. In addition, abscisic acid (ABA), a plant stress hormone, induces the closure of leaf stomata (microscopic pores involved in gas exchange), thereby reducing water loss through transpiration, and decreasing the rate of

http://dx.doi.org/10.5772/54223

photosynthesis. These responses improve the water-use efficiency of the plant on the short term [25]. The Figure 2 reveals physiological, biochemical and molecular responses to this abiotic stress in plants. It is very important highlight that most of these responses are similar in salt-stressed plants.

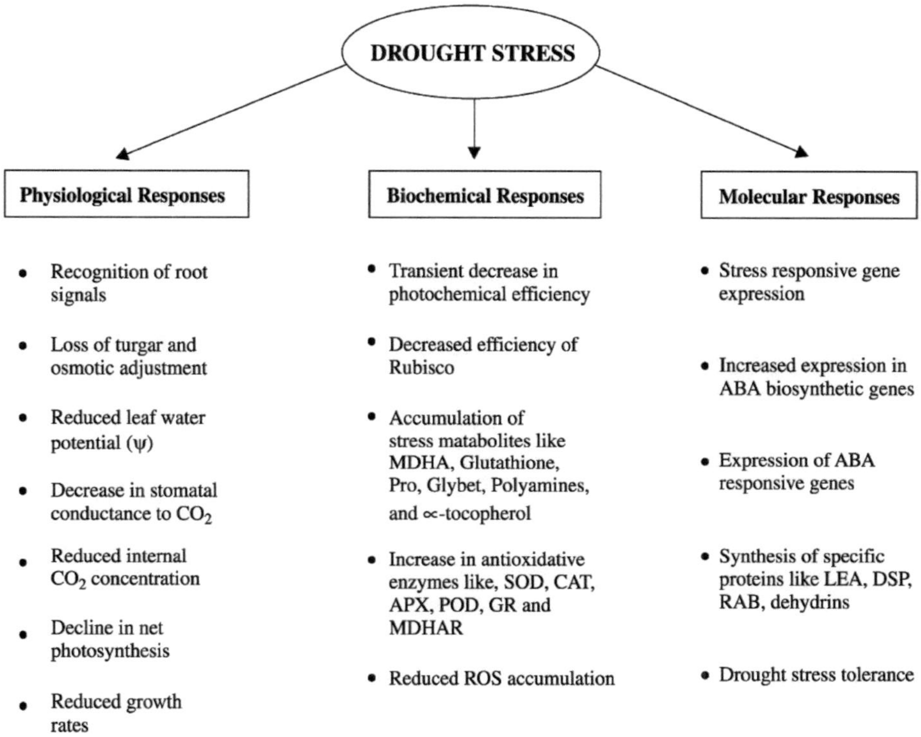

Figure 2. Physiological and molecular basis of drought stress tolerance (adapted from [8, 26]).

In salt-stressed plants, in addition to osmotic effects it is also affected by toxic damages as function of nutritional disequilibrium and high salt levels uptake for plants. Thus, salinity inhibition of plant growth is the result of osmotic and ionic effects and the different plant species have developed different mechanisms to cope with these effects [6]. The osmotic adjustment, i. e., reduction of cellular osmotic potential by net solute accumulation, has been considered an important mechanism to salt and drought tolerance in plants. This reduction in osmotic potential in salt stressed plants can be a result of inorganic ion (Na^+, Cl^-, and K^+) and compatible organic solute (soluble carbohydrates, amino acids, proline, betaines, etc) accumulations [27]. However these changes are only any few initial responses of many others occurred from salt-stressed seedlings. For instance this behavior, in the Figure 3 has been showed a schematic summary of the stresses that plants suffer under high salinity growth condition and the corresponding responses that plants use in order to survive these detrimental effects.

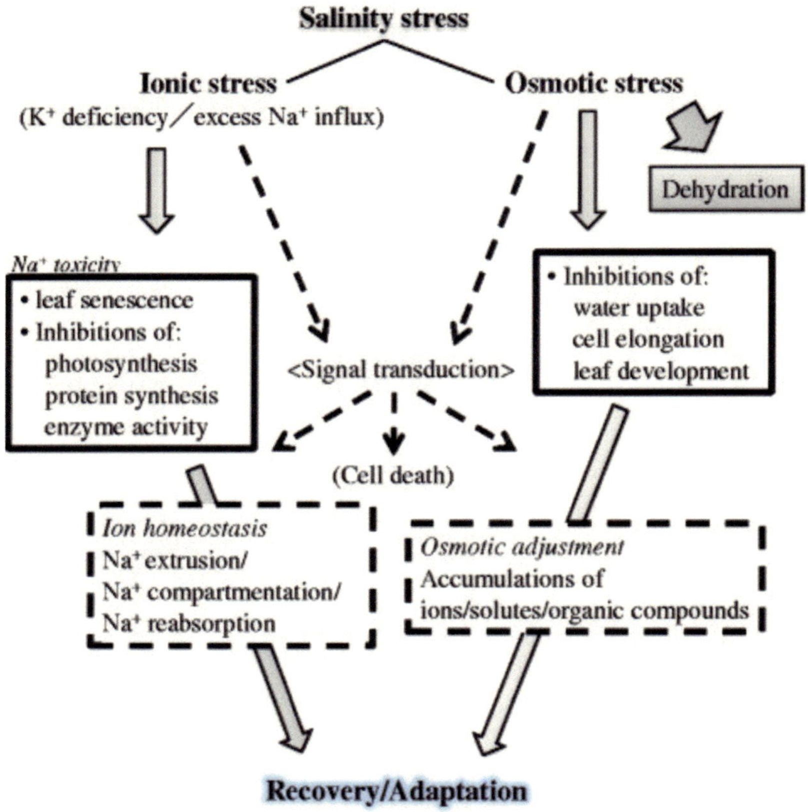

Figure 3. Adaptive responses of plants to salt stress (reproduced from [28]).

2. Water stress x salt stress effects on plant growth and development

Plant responses to drought and salinity are complexes and involve adaptive changes and/or deleterious effects. The decrease in the water potential occurred in both abiotic stresses results in reduced cell growth, root growth and shoot growth and also causes inhibition of cell expansion and reduction in cell wall synthesis [29]. According these authors, drought (likely to salinity) affects the regular metabolism of the cell such as carbon-reduction cycle, light reactions, energy charge and proton pumping and leads to the production of toxic molecules.

The literature has affirmed that plant responses to salt and water stress have much in common. For example, according reference [30], salinity reduces the ability of plants to take up water, and this quickly causes reductions in growth rate, along with a suite of metabolic changes identical to those caused by water stress. Therefore, most mechanisms were development by plants to tolerate abiotic stresses like water deficit and salinity, which are schematically showed in Figure 4.

http://dx.doi.org/10.5772/54223

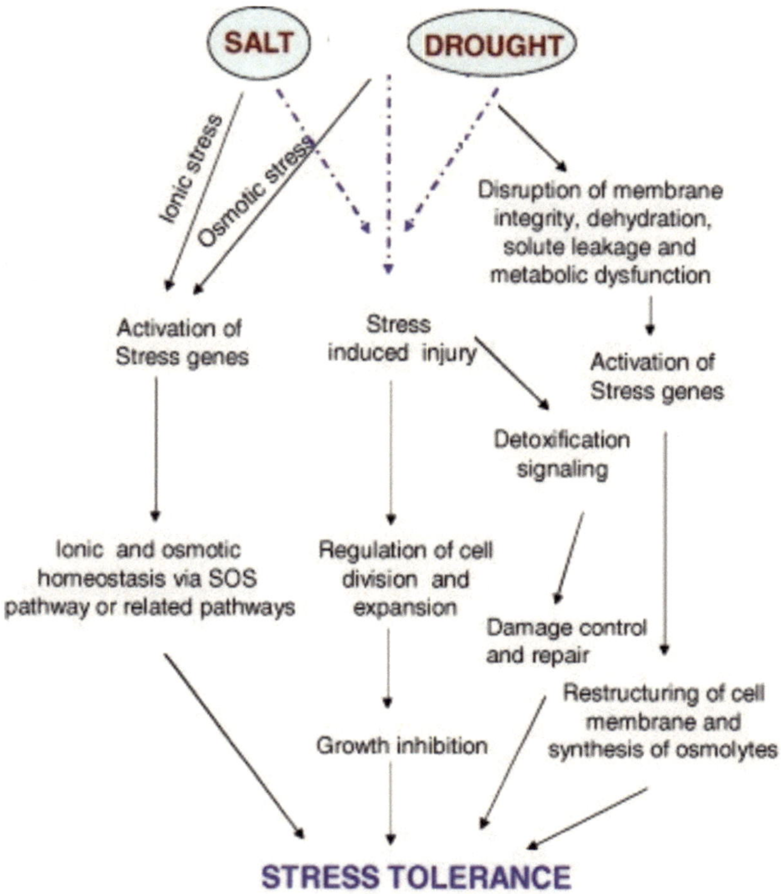

Figure 4. Water and salt stress tolerance mechanisms in plants (adapted from [31]).

2.1. Salinity and water deficit problem in arid and semi-arid regions

Arid and semiarid regions of the world such as Brazilian's Northeast, mostly faces with inadequate, irregular and erratic nature of rainfall. Moreover, according [3], in the field, drought can cause a number of plant stresses including temperature, light and nutrient stresses. However, the stress component that defines drought is a decrease in the availability of soil water. For [32], in addition with recurrent drought lack of efficient use of scarcely available water amplified the impact of water scarcity in agricultural production and productivity. These authors affirm the reduction of agricultural production results from a combination of many factors, such as crop management, crop genetics and biotic stress. Therefore, limited and irregular rainfall directly and indirectly leads to low production levels and consequently food insecurity in developing regions of the world such as Brazilian's semiarid. In addition, [33] have related it is widely recognized that land use–cover changes, such as desertification in arid and semiarid regions and deforestation in tropical zones, may exert an influence on regional or even global environmental change by changing the hydrological cycle and surface energy balance.

Looking for avoid the water stress and to improve the productivity at arid and semiarid regions generally adopt the use of irrigation, principally the drip. In this way, drip irrigation has the potential to increase crop yields with less water and it can maintain relatively constant soil water content over time near the drip lines. However, [34] have related a disadvantage of drip irrigation is that salt accumulates near the periphery of the wetted area. According them, this salt accumulation can be a matter of concern if the emitter placement does not coincide reasonably well with the location of the plant row, particularly for crops that are sensitive or moderately sensitive to soil salinity.

2.2. Salinity in soil is increased by the water deficit and aridity conditions

In arid regions, there may be periods of no net downward percolation and hence no effective leaching, so salts can accumulate in the soil. In reference [19], the combined effect of meager rainfall, high evaporation, the presence of salt-bearing sediments, and – in many places, particularly river valleys and other low-lying areas – the occurrence of shallow, brackish groundwater which gives rise to saline soils. In addition, the irrigation has been one of the major practices that more contributes with the soil salinization, like showed in Figure 5.

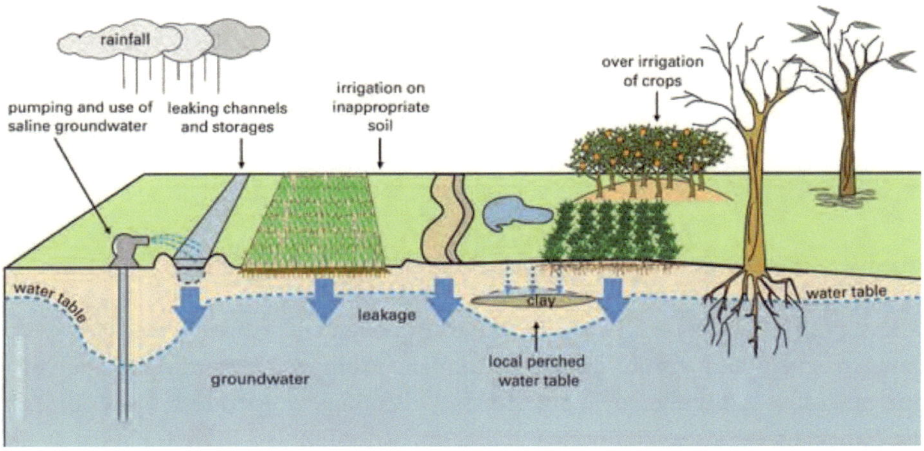

Figure 5. Causes of irrigation salinity (reproduced from [35]).

Less obvious than the appearance of naturally saline soils, but perhaps more insidious, is the inadvertent induced salination of originally productive soils, caused by human intervention. Irrigation waters generally contain appreciable quantities of salts. (For example: even with relatively good-quality irrigation water containing no more than 0.3 kg salts m3, applying 10,000 mm water per season would add as much as 3 000 kg salts ha^{-1} per year!) Crop plants normally extract water from the soil while leaving most of the salt behind. Unless leached away (continuously or intermittently), such salts sooner or later begin to hinder crop growth [19].

These deleterious effects previously showed caused by salinization can be managed by changed farm management practices. In [30] has been indicated in irrigated agriculture, better irrigation practices, such as drip irrigation, to optimize use of water can be employed. In rain-fed agriculture, this researcher suggest practices such as rotation of annual crops with deep-rooted perennial species may restore the balance between rainfall and water use, thus preventing rising water tables bringing salts to the surface. All such practices will rely on a high degree of salt tolerance, not only of the perennial species used to lower a saline water table, but also of the crops to follow, as some salt will remain in the soil.

2.3. Salt stress and irrigation and bad water uses

Irrigation water quality can have a profound impact on crop production. In reference [36], the work affirms that all irrigation water contains dissolved mineral salts, but the levels and composition of the salts vary depending on the source of the irrigation water. Salinity from irrigation can occur over time wherever irrigation occurs, since almost all water (even natural rainfall) contains some dissolved salts. When the plants use the water, the salts are left behind in the soil and eventually begin to accumulate. Since soil salinity makes it more difficult for plants to absorb soil moisture, these salts must be leached out of the plant root zone by applying additional water. This water in excess of plant needs is called the leaching fraction [37].

Salination from irrigation water is also greatly increased by poor drainage and use of saline water for irrigating agricultural crops. Therefore, inefficient or bad irrigation water and drainage systems are a major cause of excess leakage and increase the risk of salinity and waterlogging in irrigation areas. Poor water distribution on paddocks results in some areas being under-irrigated, causing salts to accumulate (where watertables are high) and other areas being over-irrigated and waterlogged. Groundwater mounds can develop under irrigation areas as a result of leakage from inefficient systems and restrictive layers. This puts pressure on the regional groundwater system forcing saline groundwater into waterways. Irrigating with saline water adds salt to the soil and increases the need for applying more irrigation water to leach salts past the plant root zone [38].

From studies of FAO [39] we can affirm that prevention and reclamation of salt-affected soils require an integrated management approach, including consideration of socioeconomic aspects, monitoring & maintenance of irrigation schemes and reuse and/or safe disposal of drainage water. Implementation of efficient irrigation and drainage systems and good farming practices can prevent and, in some cases, reverse salinization. If appropriate management practices are not applied in time, it may be necessary to take the land out of production altogether. Moreover, actions to fight or mitigate salinization can be implemented by local institutions and research stations, while research and technology transfer can play a crucial role in providing tools, setting up management strategies or spreading water-saving techniques.

2.4. Growth and plant development affected by multiple stresses as water and salt stress

Plant growth can be limited by water deficit and by excess water. Water deficit occurs in most natural and agricultural habitats and is caused mainly by intermittent to continuous periods without precipitation. Excess water occurs as the result of flooding or soil compaction [12].

When plant cells experience water deficit, cellular dehydration occurs, this promotes reductions of cell turgor (Ψ_p) and cell volume. It is also associated with the water potential (Ψ_w) of the apoplast becoming more negative than that of the symplast. Cellular dehydration could causes ion concentration reaching cytotoxic levels [12].

Water deficit not only decreases turgor pressure, but also decreases wall extensibility and increase yield threshold. The water deficit effects on yield threshold are presumably involved in complex structural changes of the cell wall that may not be readily reversed after relief of water deficit [12].

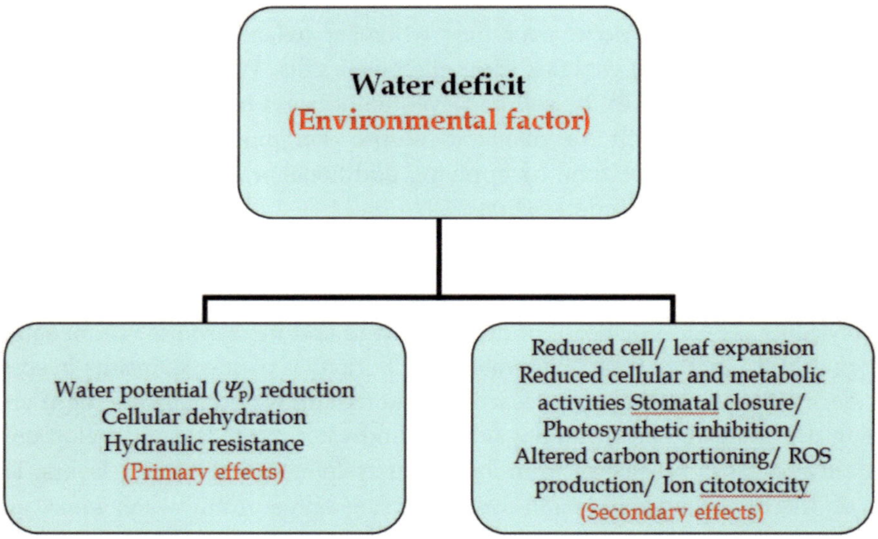

Figure 6. Some physiological and biochemical perturbations in plants caused by fluctuations in abiotic environments [12].

Salinity can have similar aspects to water stress in plant growth, except for the addition of ion cytotoxicity, which appears with salt excess in soil. [12].

The toxicity of high Na^+ and Cl^- in the cytosol is due to their specific ion effects. High salt concentrations, outside the cell, can result in osmotic stress. Once in the cytosol, however, certain ions act specifically, either singly or in combination, affecting whole plant because ions move to the shoot in the transpiration stream [6].

The increase of solutes on root medium, mainly the ions, can promote a reduction of water absorption by plant root system, contributing to reduction of root conductivity. As consequence, the plant absorbs less water, and if the transpiration rate is higher than water absorption rate, and the result is a water deficit, which could result in photosynthetic rate reduction and growth rate reduction [40]. According to the figure below, salt stress effects in timescale alters initially, absorption of water and nutrients, and membranes permeability [41].

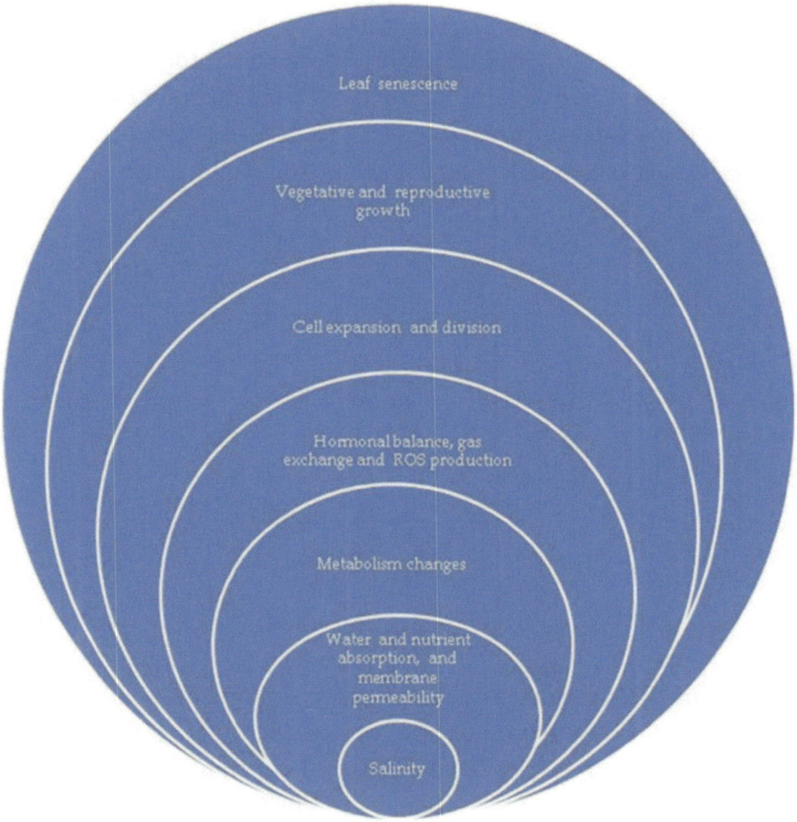

Figure 7. Change in plant physiology as consequence to salinity adapted from [40].

These previous changes reflect in nutritional and hydric balance of plants and promote changes in metabolism, in hormonal balance, gas exchange and ROS production. All these changes undertake cell expansion and division, vegetative and reproductive growth and acceleration of leaf senescence, which result in plant death [12].

An important strategy to avoid ion toxicity is the accumulation of ions during osmotic adjustment in vacuoles, where the ions are kept out of contact with cytosolic enzymes or organelles. Many halophytes utilize vacuolar compartmentalization of Na^+ and Cl^- to facilite osmotic adjustment that sustain or enhances growth in saline environments. When ions are compartmentalized in vacuoles, at the same time other solutes must accumulate in the cytoplasm to maintain water potential equilibrium within the cell, as example of these compatible solutes, can be cited the quaternary ammonium compounds, sugar alcohols, quaternary ammonium compounds (QACs) and tertiary sulfonium compounds (TSCs) [12].

According to [6], the growth responses to salinity stress occurs in two phases: a rapid and intense response to the increase in external osmotic pressure related to NaCl increase in the medium, which contribute to a stronger reduction in growth. Followed, it can be observed, a slower response due to toxic ion accumulation in tissues (Ionic phase), which is related to a severe toxicity in leaves, represented in most of times by chlorosis and can be originated necrosis in these tissues (See figure 8). Leaf injury and death is probably due to the high salt load in the leaf that exceeds the capacity of salt compartmentation in the vacuoles, causing salt to build up in the cytoplasm to toxic leaves [6,30]. These responses vary strongly between genotypes, salinity levels, soil and other abiotic factors [6].

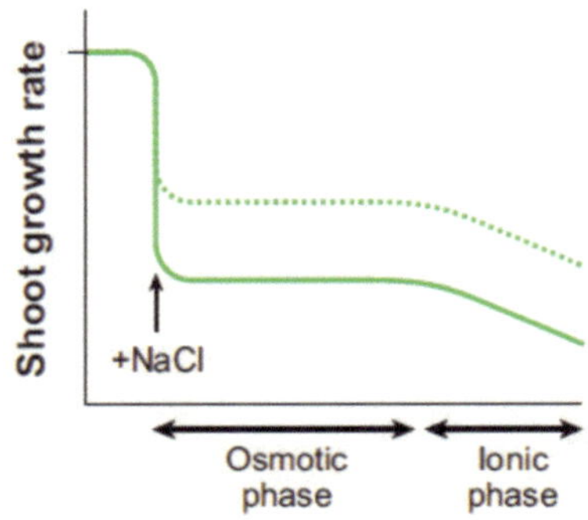

Figure 8. Growth responses to salt stress from [6].

Plants undergo characteristic changes from the time salinity stress is imposed until they reach maturity [30]. Salinity inhibition of plant growth is the result of osmotic and ionic effects and the different plant species have developed different mechanisms to cope with these effects [30]. The plant responses to salinity are different between the species, the salt levels and also depends of the time exposition. In the table below could be observed the plant response to salinity at different time scales [Table modified from 30].

	Water stress effects	Salt-specific effects
Time	(Observed effect on growth of a salt-tolerant plant)	(Additional effects on growth of a salt-sensitive plant)
Minutes	Instant reduction in leaf and root elongation rate then rapid partial recovery	
Hours	Steady but reduced rate of leaf and root elongation	
Days	Leaf growth more affected than root growth; Reduced rate of leaf emergence	Injury visible in oldest leaf
Weeks	Reduced final leaf size and/or number of lateral shoots	Death of older leaves
Months	Altered flowering time, reduced seed production	Younger leaves dead, plant may die before seed matures

Table 1. Water and salt-specific effects on plant growth

As can be observed in this table, in the first few seconds or minutes, it has been observed that cells lose water or shrink. When the hours are passing, cells regain the original volume but cell elongation is reduced, contributing to lower rates of leaf and root growth. After some days, cell elongation and cell division promote alterations in leaf size appearance [30]. In plants, with high salt absorption, the oldest plants become senescent, and some leaves can dead. After months, differences between months, with high and low salt uptake become very pronounced, with a large amount o leaf injury and complete death in some cases if the salinity level is high enough [30].

The tolerance or plant response to salt can vary with cultivars, the plant age and growth stage, environmental conditions, cultural practices, irrigation management, soil fertility, and the intensity or other plant stress. Another, stress factor is wind that can be injurious to plants as salt. When the two are combined near the ocean, plants suffer even greater damage. Some crassulaceae and succulent plants are highly salt and wind tolerant, in different way, moderate salt tolerant plants of soil salinity usually tolerate light salt spray but should not be used in exposed locations [42].

The comparative effect of salinity and water stress on growth was analyzed in *Jatropha curcas* seedlings. The plants subject to water stress, through the exposition to PEG, showed higher restrictions in leaf growth compared to the salt-stressed ones. Moreover, only the PEG treatment caused a pronounced effect on leaf membrane integrity. Since the plants subjected to mild water and salt stresses showed a rapid and almost complete recovery, these physiological alterations could represent a set of adaptive mechanisms employed by *J. curcas* to cope with these stressful conditions [43].

Pinus seed germination was evaluated in different conditions: water stress simulated by PEG agent experiments, salt stress using sodium chloride (NaCl) and diluted sea water, associated to different temperatures (heat stress). Seed germination and the activities of the main enzymes involved in Pinus seed reserve utilization (glyoxylate cycle enzymes) decreased with increasing PEG, NaCl and seawater concentrations. In addition, the seawater treatment was the less severe on seed germination. As regard to heat treatments, the maximum germination percentage (80%) was obtained at 80°C and short exposure time (3 min) [44].

Another study involving water and salt stress effects on plant growth was verified in Tamarisk (*Tamarix chinensis* Lour.) seedlings, where it was observed that the water supply and salt content had interactive effects on Tamarisk biomass, while interactive effects of the both stress had no influence in biomass allocation. This parameter was more affected by water supply, which implied that the balance between soil water availability and atmospheric evaporative demand defined the biomass allocation [45].

Salinity affects both vegetative and reproductive plant development, which promotes negative implications depending on the harvested organ, stem, leaf, root, shoot, fruit, fiber or grain [46]. Salinity generally reduces shoot growth more than root growth and can reduce the flowering and increase sterility. Considering the salt-tolerance importance from an agronomic or horticulturist perspective is based on the yield of the harvestable organ, relative to that in non-stressed environments, understanding how salinity affects vegetative and reproductive development is important for developing management strategies that can minimize stress at critical times [46]. Salinity stress could delay germination, although most plants are tolerant and there may be no difference in the percentage of germinated seeds.

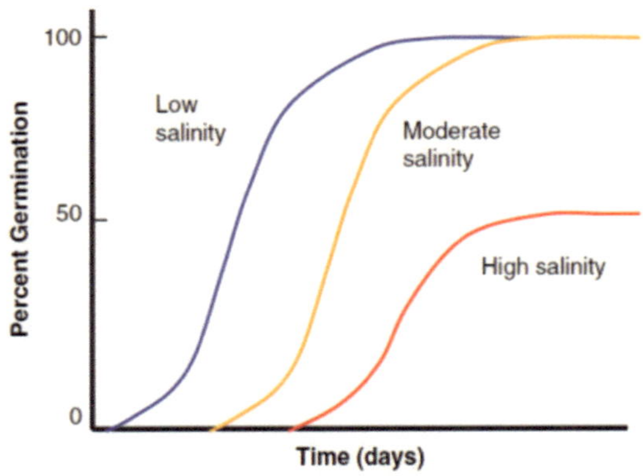

Figure 9. Salinity effects on relationship between percent germination and time after water addition at low, moderate and high salinity, modified from [46].

Helianthus annuus seed germination was evaluated using treatment to overcome salt and drought stress during germination. In this study, it was verified that seeds subjected to salinity and water stress showed reduced germination in both conditions [47]. Germination, root and shoot length were higher in NaCl than PEG conditions, but mean germination time and abnormal germination percentage were lower in these conditions [47]. The authors concluded that germination inhibition at the same water potential of NaCl and PEG resulted from osmotic effect rather than salt toxicity, in addition it was observed that hydropriming increased germination and seedling growth under salt and drought stresses [47]. Another study with germination approached salt and water stress on seed germination and early embryo growth in two cultivars of sweet sorghum. In this study, it was observed that salt stress prolonged the germination time. Shoot and root growth was adversely affected by water stress (PEG), whereas the effect of salt stress was less relevant [48].

2.5. Salt tolerant plants "Halophytes" and water stress tolerant plants

Halophytes are plants that can survive and reproduce in environments with high salt concentration (200 mM NaCl), these plants constitute about 1% of the world's flora. Halophytes can be classified as "natural" and plants that tolerate salt but do not normally live in saline conditions [49]. Other classifications of halophytes are based on the characteristics of naturally saline habitats [49] or the chemical composition of the shoots ('physiotypes', [50] or the ability to secrete ions (recreto-halophytes, [51]).

The effect of salinity on growth varies amongst halophytes, which can be observed in figure bellow for different species [49]. According this figure, *Sueada maritima* was the specie that could growth in the highest NaCl concentration, however the best condition was around 40 mM NaCl. The most affected specie was *Thellungiella halophila,* which showed stronger growth reduction from 200 mM NaCl. In addition, it can be highlighted that dicotyledonous species showed higher tolerance to NaCl than monocotyledonous species, except for *T. halophila* [49].

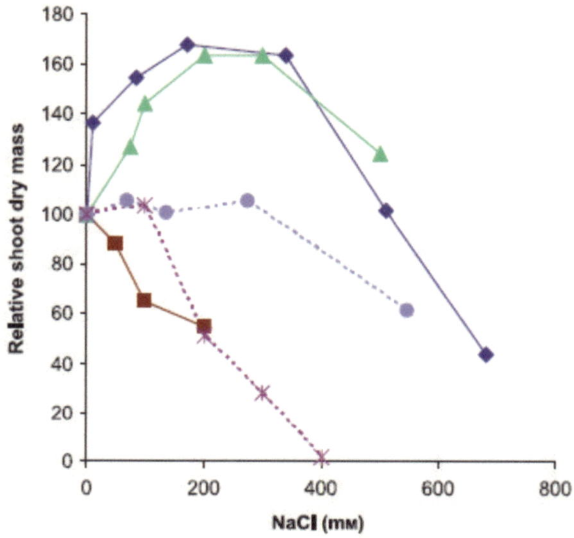

Figure 10. Effect of NaCl increasing on the growth on shoot dry mass in different halophytes. Solid lines represents dicotyledonous species; broken lines, monocotyledonous species. (◆)*Suaeda maritima* (35 d), (■) *Thellungiella halophila* (14 d); (▲) *Disphyme australe* (60 d); (*) Puccinellia peisonis (42 d); (●) Distichlis spicata (21 d). (Modified from 49)

The salt requirement by halophyte during their growth and development were evaluated in two halophytic species (*Inula crithmoides* L. and *Plantago crassifolia* Forssk.) and a maritime dune species (*Medicago marina* L.) [52]. In this study was observed that growth parameters results showed that all three species grew much better on the salt-free and nutrient-rich substrates, peat and garden soil, than on saline soil and dune sand. These results indicate that salts are not necessarily required for development of halophytic species, and suggest that limitation of water and nutrients, rather than soil salinity per se, are the most important restrictive factors for plant growth in saline habitat [52].

Studies involving extreme dicotyledonous halophytes show that optimal growth in the presence of low or moderate salt concentrations; on the other hand, this stimulatory effect is not commonly detected in monocotyledonous halophytes, nor in other salt-tolerant dicots, which grow best in the absence of salt [53]. The salt tolerance degree usually varies among halophytes and, for a specific species, also at different developmental stages [53], but inhibition of plant growth is always observed at sufficiently high salinity levels, in all investigated taxa [52, 53].

Tamarix chinensis that is used as a windbreak for preventing desertification and erosion in arid areas and potential bio-energy resources for marginal land utilization [54], was evaluated with relation to salt and water stress on biomass and photosynthetic characteristic. In this study was observed that soil salt increment had no significant effect on biomass allocation of *T. chinensis* seedlings [45], while water deficit significantly favoured root biomass accumulation [45].

In study about growth parameters and anatomical changes in the halophyte *Atriplex nummularia* Lindl, it was observed that the height, diameter and fresh and leaf dry biomass, stems and roots were sensitive to soil moisture. Moreover, anatomical changes in the vesicular trichomes of the leaves were detected with water soil variations and druse formed by deposition of sodium chloride within cells [Figure from 55]. A low water content in the oil is related to small average diameters and volumes of the epidermal vesicles [55].

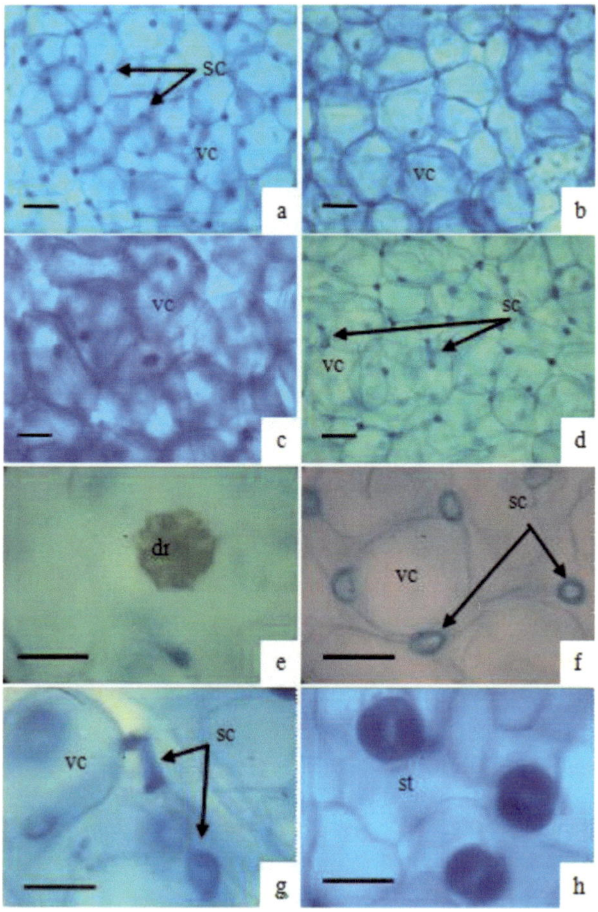

Figure 11. Epidermis front view of the leaves of Atriplex nummularia Lindl. grown in sodic saline soil under different water regimes: (a) 35%; (b) 55%; (c) 75; (d) 95% of FC. Vesicular trichome consisting of stalk cell surmounted by a vesicular cell (f–g) and stomata (h). sc: stalk cell; vc: vesicular cell; dr: drusa; st: stoma. Bar: 50 µm (a–d), 25 µm (e–h). Modified from [55]

In another study also with A. *nummularia* was observed that leaves display a most prominent osmotic adjustment than roots. In addition, it has been also concluded that Na^+ and Cl– ions are the major components to the overall osmotic adjustment, in both salt-untreated and treated leaves but in untreated roots the K^+ is the major component [56]. For *Atriplex allimus*, the decrease in fresh weight might be attributed to the decrease in water content. Moreover, the decrease in photosynthesis could result from a decrease in stomatal conductance, which corresponds to a protective mechanism against water loss to improve water use efficiency [57]. The *A. allimus* could tolerate NaCl and drought through decreasing growth, reducing gas exchange parameters to improve water use efficiency. In another study with halophytes, it was verified the effect of salinity and soil drying on nutrient uptake and growth of *Spartina alterniflora* in a simulated tidal system [58]. This study demonstrated that high salinity associated with drought decreased survival to 71%, Furthermore, leaf chlorosis and browning occurred in plants subjected to increased salinity and drought conditions, which can demonstrated that drought may magnify the adverse effects of salinity on plant nutrient status [58]. Some plants can develop a lot of mechanisms to support water deficit, as example the development of internal water deficit, which can be important for some crops such as coffee and mango in order to trigger phenological events such as flower bud release. Additionally, it has been know that for a tropical environment the range of evaporative demand on average is higher than that of temperate zones. This implies that leaf water status changes much more diurnally in tropical trees than in many temperate trees or annuals, and leaf water deficits may occur under the high evaporative demand even without any soil water shortage, such as in banana, cocoa, coffee, papaya, and tea [59].

The development of thick cuticular membranes generally is interpreted as an adaptation to drought with regard to the formation of an efficient transpiration barrier. The capacity of *Teucrium chamaedrys* in maintaining a low but still measurable stomatal conductance under drought stress at maximum light intensity can be interpreted as a mechanism to adjust leaf temperature below the heat resistance limit by transpiration cooling [60].

Jatropha curcas L. is a biodiesel crop that is resistant to drought stress, and the tolerance of *Jatropha curcas* seedlings to NaCl was evaluated through some ecophysiological analyses. This study showed that the negative NaCl effects on *J. curcas* crop, are mainly due to Cl- and /or Na+ toxicity and to a nutritional imbalance caused by increase in Na^+/K^+. Moreover, the osmotic effect on salinity in this species is very pronounced, probably due to strong control of leaf transpiration, which reduces water loss [61].

The NaCl differential tolerance of different maize genotypes were evaluated through some growth and physiological parameters and it was concluded that the SDM/RDM ratios, leaf Na+ content or leaf soluble organic solute content had no relation with salt tolerance. On the other hand, Na^+ and soluble organic solute accumulation in roots due to salinity appeared to play an important role in the acclimation of maize genotypes, being that these characteristics could be used as physiological markers to salinity [62].

2.6. Studies involving adaptations or acclimations of tropical plants to water and and salt stress

Plants under salt stress can occasionally tolerate alterations in the environmental conditions. Salt tolerance in plants is determinate by multiple biological traits that will determine their water retentions and/ or acquisition capacity in safeguarding photosynthetic functions and ion homeostasis. The plant ability of eliminating free radical under salinity and water stress using active osmolytes, usually demand a lot of energy. Many plant species that are salt tolerant to salt stress show the ability to produce and accumulate osmoprotectants [63].

Water deficit and saline soils of the Brazilian semi-arid northeastern region are limiting factors and most of the times negatively affect the plant development. *Mimosa tenuiflora* popular known as Jurema preta is a small, multiple use tree that is widely distributed in environments with severe water stress. In this study, it was verified that Jurema preta seeds germination showed more sensitive to NaCl than to water stress, and due to the stronger germination reduction in this condition, this species can be classified as a glycophyte [64].

Some studies evaluate the plant acclimation to abiotic stresses, and it is very common the pretreatment use for alleviates the negative effects on plants. The H_2O_2 pretreatment effect in maize seeds during germination and seedling acclimation to salinity, and it was observed that H_2O_2 pretreatment of seeds induced acclimation of the plants to salinity. It decreased the salinity deleterious effects on the maize growth. In addition, it was also verified differences in antioxidative enzyme activities, which may explain the increased tolerance to salt stress of plants originated from H_2O_2 pretreated seeds [65]. Another plant that is cited as water tolerant is *Jatropha curcas*, which was evaluated with relation to organic and inorganic solutes contribution of to osmotic adjustment under salinity. This plant showed an effective osmotic adjustment in leaves under salinity, maintaining their hydration status, mainly via Na^+ and Cl^- accumulation, and in relation to organic solutes, glycinebetaine was more important to osmotic adjustment than proline in both salt-treated and untreated plants [66]. The salt tolerance was evaluated in cotton, sorghum and maize, which are considered very adapted to Brazilian semi-arid and show contrasting tolerance to salt [67]. In this study, it was observed that growth results confirmed the highest salt tolerance to cotton, whereas cowpea presented highest susceptibility to salinity, because it was verified stronger reduction in shoot and root dry mass from moderate salinity level. In addition, antioxidant enzyme activities changes suggest that cotton constitutive enzyme system seems to be more efficient than the others [67]. Cowpea is widely cultivated in arid and semi-arid regions of the world where salinity is a major environmental stress that limits crop productivity [68]. This author evaluated the relationship between salt tolerance with differences in ion accumulation, biomass allocation and photosynthesis in cowpea cultivars. Salt stress caused reduction in biomass accumulation and photosynthetic rates per unit leaf mass, mainly in 'TVu' cowpea cultivar. 'Pitiúba' showed the relative advantage over 'TVu' cultivar under salt stress, which was associated with reduced Na^+ accumulation in leaves related to increase in Na+ concentration in roots at early stages of salt treatment. In general way, these responses would allow greater whole-plant carbon gain, thus contributing to a better agronomic performance of salt-tolerant cowpea cultivars in salinity-prone regions [68]. In another study, seven cowpea cultivars were analyzed in relation to growth and organic and inorganic contents in response to salinity. It was demonstrated that salinity reduced dry mass of all cultivars, and 'Pitiúba' and 'Vita 5' were the most tolerant and 'TVu' the most sensitive [69]. The growth parameters showed strong correlation with the salt tolerance degree but increased significantly with the salt increment concentration only in the most tolerant cultivars. Overall, Na^+ and Cl^- shoot content increased due to salinity but accumulated drastically only the more susceptible cultivars, particularly in 'TVu' cowpea cultivar. K^+ content in leaves was higher than Na^+ and Cl^- contents but none of the ions were significantly in this tissue. Consequently the Na^+/K^+ ratio was higher in the most sensible cultivars. This may represent an important factor for the growth reduction observed in the stressed plants. Considering the organic solutes (proline, soluble carbohydrate and N-amino solutes) were not correlated with NaCl degree tolerance in the cultivars [69]. Another species that is widely distributed in Brazilian semi-arid region is cashew nut (*Annacardium occidentale* L.) showing some adaptations and acclimations to water deficit, and it has been strongly used as a model to evaluate salt tolerance mechanisms. In relation to seed germination and seedling establishment of dwarf-cashew, salinity delayed emergence and establishment seedling, however, only the final percentage of established seedlings was reduced, accompanied by inhibition of reserves depletion and an embryonic axis dry mass decrease. [70]. The salinity effects on seed germination and seedling establishment could also be resulting of negative salt influence on metabolic processes, such as the mobilization of the stored reserves for the growing embryonic axis. In addition, salinity could inhibit seedling development through the inhibition of enzymes involved in the mobilization of lipid reserves of the seed [70, 71]. Also with relation to cashew nuts, it was evaluated the changes in physiological indicators in two contrasting cashew rootstocks. Among the physiological parameters evaluated, leaf membrane damage was closely associated with the differences in salt tolerance between the two cashew genotypes. In addition, under salinity the tolerant rootstock showed greater ability to accumulate compatible organic solutes (amino acids, proline and soluble sugars) in leaves in addition to maintaining the soluble sugar concentration in roots as compared with the sensitive rootstock [72]. Further, evaluating physiological and biochemical changes occurring in dwarf-cashew seedling under salinity showed that organic solutes was few affected by salt stress, except for proline that showed in increment. In addition, the most representative enzyme activity in roots was guaiacol and ascorbate peroxidase in removing reactive oxygen species, due to roots were considered more protected to salinity stress than leaves. In addition, changes in protein pattern were related mainly to salt stress rather than by alteration in developmental stage [73]. Another study with salt-stressed cashew nuts suggest that the ability of these seedlings to adapt to salt stress is, at least in part, correlated to maintenance of integrity and protection against oxidative plasma membrane damage, which could contribute to the activation of H^+-ATPase plasma membrane, as a cellular mechanism to regulate ion exclusion from the shoot [74]. In another study, it was evaluated the gas exchange in leaves of sun and shade of cashew in different water regimes, where it was verified that light exposition influenced more stronger the physiological parameters than suppression of water. It occurred probably because the presented results were better related to the phenological stages than with the treatments, demonstrating a conspicuous physiological effect of the strength of the drains and the irradiance in these plants [75]. Also studying adult plants of cashew, it was verified that the increase in electric conductivity did not contributed to alterations in plant growth and development, once it was not observed changes in evaluated parameters. Metabolic alterations was more related to plant developmental stage, showing higher values in fructification stage [76].

The water relations in six adult species of Caatinga, a typical vegetation from Brazilian semi-arid, in the middle of the dry season. Based on results, the trees were classified into four groups: (I) *Mimosa caesalpiniifolia* had low leaf water potential (Ψw) at predawn and no significant decrease at midday; (II) *Caesalpinia pyramidalis* and *Auxemma oncocalyx* had low Ψw at predawn and significant decrease at midday. For these species the recuperation of water status at night may have been sufficient for maintaining stomata open during the day; (III) *Caesalpinia ferrea* and *Calliandra spinosa* had relatively high Ψw at predawn and a significant decrease at midday. They might also have deeper root systems; and (IV) *Tabebuia caraiba* with the highest Ψw at predawn and no significant decrease at midday, possibly indicating a combination of good stomatal control of water loss and a deeper root system [77].

In relation to commom bean plants, mild water deficit affected the photochemical apparatus in these plants, probably due to by down-regulation since plants did not show photoinhibition. The photochemical apparatus of A222 and A320 genotypes was more sensitive to drought stress. On the other hand, even after 10 d of water withholding, the maximum efficiency of photosystem 2 was not affected, what suggest efficiency of the photoprotection mechanisms [78].

The seasonal variations of physiological aspects of caatinga species demonstrated the high levels of water potential, even in the dry period, a situation in which commonly occurs water stress due to absence of soil water. It is related to water potential decrease during the absence of water in soil. The fluorescence data showed that the photosynthetic states were good with no apparent deficiency of water. In addition, it was verified that the survival strategies of these species to water deficit are efficient and result from a highly complex evolution [79].

In a study with *Vigna unguiculata*, an important crop largely cultivated in warm and hot lands, it were evaluated some photosynthetical parameters and metabolic changes in cowpea under water stress and recovery treatments. In this study it was verified that decreased CO_2 assimilation rates during water stress were largely dependent on stomatal closure, which reduced available internal CO_2 and restricted water loss through transpiration [80]. Additionally, the alterations in carbohydrate metabolism showed an accumulation of soluble sugars in water-stressed leaves, which also persisted in rewatering plants for 1 day. The authors conclude that a transient end-product inhibition of photosynthesis, contributing to a minor non-stomatal limitation during stress and recovery initial phase [80].

In relation to *Phaseolus vulgaris,* the mild water stress effects on photosynthetic parameters and water potential were analyzed in five common bean. The stomatal conductance and net photosynthetic rate were significantly reduced in all genotypes due to the water deficit [81]. Mild water deficit affected the photochemical apparatus in bean genotypes probably by down-regulation since plants did not show photoinhibition. It was also interesting in this study, that even after 10 d of water withholding, the maximum efficiency of photosystem 2 was not affected, what suggest efficiency of the photoprotection mechanisms [81].

In the figure below, it is resumed long-term or acclimations responses to drought stress and short-term responses that reach the plant perceive the water stress and develop the following responses against this condition [82]. The majority of traits related to plant drought adaptations, as example root size and depth, hydraulic conductivity and seed storage reserves are associated to plant development and structure and are constitutive rather than stress induced. In addition, a large part of drought plants resistance to drought and its ability to get rid of to excess radiation. The natural mechanisms responsible to leaf photoprotection, especially to related to thermal dissipation and oxidative stress. Plants can also endure water drought conditions by avoiding tissue dehydration, while maintaining tissue water potential as high as possible, or tolerating lower water potential. Dehydration avoidance is usually observed in annual and perennial plants, which has been associated to proper characteristics as example: capacity to minimize water loss and maximize water uptake [82]. The water loss could be done by stomata closer, reducing light absorbance by leaf or through reflectance increasing due to the development of a trichome layer, or by decresing of leaf surface with reduced growth or by shedding of young by old leaves. On the other hand, the maximizing water could be reaching by increasing in root system, which is resulting of alteration in allocation of plant sources [82].

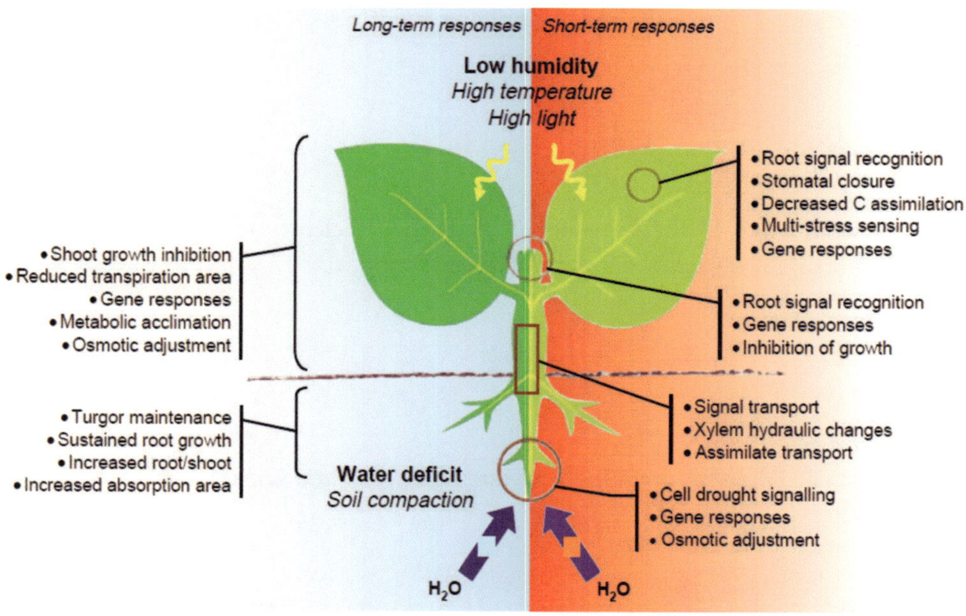

Figure 12. Plant responses to water stress from 82

3. Conclusion

Among the great number of abiotic stress affecting plants, drought and salinity are the most severe and stronger ones that limit plant growth and crop productivity in agriculture worldwide. These stress could have damages exceeding the sum of that attributed to all other natural disasters, and when they happen together promoted devastating changes in plants subjected to them. Plant responses to drought and salt stresses have much in common, the water limitations in water stress is very difficult to plant tolerate and maintain its normal growth and development. Salt stress occurs from both osmotic stress due to low water potentials and salt-specific effects, for this beyond the water restrictions, salt-stressed plants have to develop conditions to tolerate the toxic effects caused by ion accumulation, which could affect all aspects of plant metabolism. In the first view, we could speculate that salt stress in plants are more severe than water stress, by the fact that the first condition show two components, however, it is early to conclude this. This is because studies on the comparative physiology of plants to water and salt stress are few, and it is necessary perform more studies involving different plant species, as cultivate and as native, subjected at the same time to both stress to conclude what is the most severe for plant growth and development. Moreover, the development of new tools and strategies to evaluate the combination of water and salt stress on plant are also necessary, but the perspectives in relation to the better understanding of how some plants could tolerate, escape, acclimate and adaptate to these severe abiotic stresses. This knowledge could have the support of studies involving since molecular (genomics and proteomics tools) aspects as ecotype researches (isotopes, imagging) and this combination will allow the faster comprehension of plant perfomance in different environments, as well the brinding of genotype and phenotype gap.

Acknowledgements

The authors would like to thank National Institute of Science and Technology Salinity (CNPq/MCT/Brazil) for the fellowships and financial support.

Author details

Alexandre Bosco de Oliveira[1], Nara Lídia Mendes Alencar[1,2] and Enéas Gomes-Filho[1,3]

1 Federal University of Ceará, Brazil

2 Federal Institute of Science and Technology of Ceará/Campus Crateús, Brazil

3 National Institute of Science and Technology Salinity/CNPq, Brazil

References

[1] Araus JL, Slafer GA, Reynolds MP, Royo C. Plant breeding and drought in C-3 cereals: what should we breed for? Annals of Botany 2002;89(7) 925-940.

[2] Tester M, Bacic A. Abiotic stress tolerance in grasses. From model plants to crop plants. Plant Physiology 2005;137(3) 791–793.

[3] Verslues PE, Agarwal M, Katiyar-Agarwal S, Zhu J, Zhu JK. Methods and concepts in quantifying resistance to drought, salt and freezing, abiotic stresses that affect plant water status. The Plant Journal 2006;45(4) 523-539.

[4] Intergovernmental Panel on Climate Change (2007). http://www.ipcc.ch. Accessed 25 October 2007.

[5] Passioura JB. The drought environment: physical, biological and agricultural perspectives. Journal of Experimental Botany 2007;58(2) 113–117.

[6] Munns R, Tester M. Mechanisms of salinity tolerance. Annual Review of Plant Biology 2008;59(1) 651-681.

[7] Vinocur B, Altman A. Recent advances in engineering plant tolerance to abiotic stress: achievements and limitations. Current Opinion in Biotechnology 2005;16(2) 123-132.

[8] Shao HB, Chu LY, Jaleel CA, Zhao CX. Water-deficit stress-induced anatomical changes in higher plants. Comptes Rendus Biologies 2008;54(3) 215–225.

[9] Grime JP. Evidence for the existence of three primary strategies in plants and its relevance to ecological and evolutionary theory. The American Naturalist 1977;111(7) 1169–1194.

[10] Gaucherand S, Liancourt P, Lavorel S. Importance and intensity of competition along a fertility gradient and across species. Journal of Vegetation Science 2006;17(4) 455-464.

[11] Mittler R. Abiotic stress, the field environment and stress combination. Trends in Plant Science 2006;11(1) 11-19.

[12] Taiz L, Zeiger E. Plant Physiology. Sunderland: Sinauer Associates, 2009.

[13] Liancourt P, Callaway RM, Michalet R. Stress tolerance and competitive-response ability determine the outcome of biotic interactions. Ecology 2005;86(6) 1611-1618.

[14] Ives AR, Carpenter SR. Stability and Diversity of Ecosystems. Science 2007;317(5834) 58-62.

[15] Butterfield BJ. Effects of facilitation on community stability and dynamics: synthesis and future directions. Journal of Ecology 2009;97(6) 1192-1201.

[16] Villarreal-Barajas T., Martorell C. Species-specific disturbance tolerance, competition and positive interactions along an anthropogenic disturbance gradient. Journal of Vegetation Science 2009;20(6) 1027-1040

[17] Boyer JS. Plant productivity and environment. Science 1982;218(4571) 443-448.

[18] Amarjit KN, Kumari S, Sharma DR. In vitro selection and characterization of water-stress tolerant cultures of bell pepper. Indian Journal of Plant Physiology 2005;10(1) 14-19.

[19] Hillel D. Salinity; Management. In: Hillel D, Hatfield JH, Powlson DS, Rosenzweig C, Scow KM, Singer MJ, Sparks DL. (Eds.) Encyclopedia of Soils in the Environment. London: Elsevier/Academic Press; 2005. p.435-442.

[20] Ashraf M., Harris PJC. Potential biochemical indicators of salinity tolerance in plants. Plant Science 2004;166(1) 3-16.

[21] Evelin H, Kapoor R, Giri B. Arbuscular mycorrhizal fungi in alleviation of salt stress: a review. Annals of Botany 2009;104(7) 1263-1280.

[22] Cowan IR. Water use and optimization of carbon assimilation. In: Lang OL, Nobel PS, Osmond CB, Ziegler H. (Eds.) Physiological plant ecology. II. Water relations and carbon assimilation. Berlin: Springer-Verlag; 1982. p.589–613.

[23] Davies WJ, Zhang J. Root signals and the regulation of growth and development of plants in drying soil. Annual Review of Plant Physiology and Plant Molecular Biology 1991; 42(1) 55–76.

[24] Zhang J, Jia W, Yang J, Ismail AM. Role of ABA in integrating plant responses to drought and salt stresses. Field Crops Research 2006; 97(1) p.111-119.

[25] Waseem M, Ali A, Tahir M, Nadeem MA, Ayub M, Tanveer A, Ahmad R, Hussain M. Mechanism of drought tolerance in plant and its management through different methods. Continental Journal of Agricultural Science 2011;5(1) 10-25.

[26] Reddy AR, Chiatanya KV, Vivekanandan M. Drought induced responses of photosynthesis and antioxidant metabolism in higher plants. Journal of Plant Physiology 2004;161(11) 1189–1202.

[27] Hasegawa PM, Bressan RA, Zhu JK, Bohnert HJ. Plant cellular and molecular responses to high salinity. Annual Review of Plant Physiology and Molecular Biology 2000;51(1) 463-499.

[28] Horie T, Karahara I, Katsuhara M. Salinity tolerance mechanisms in glycophytes: An overview with the central focus on rice plants. The Rice Journal 2012;5(1) 1-18.

[29] Chaitanya KY, Sundar D, Jutur PP, Reddy AR. Water stress effects on photosynthesis in different mulberry cultivars. Plant Growth Regulation 2003;40(1) 75-80.

[30] Munns R. Comparative physiology of salt and water stress. Plant, Cell and Environment 2002;25(2) 239–250.

[31] Mahajan S, Tuteja N. Cold, salinity and drought stresses: An overview. Archives of Biochemistry and Biophysics 2005;444(2) 139-158.

[32] Yihun Y, Shultz B, Haile AM, Erkossa T. Agricultural Productivity Optimization in water scarce Semi-Arid Region of Ethiopia. Delft: UNESCO-IHE; 2010.

[33] Wang Q,

[34] Takahashi H. A Land surface water deficit model for an arid and semiarid region: Impact of desertification on the water deficit status in the Loess Plateau, China. Journal of Climate 1999;12(1) 244-257.

[35] Hanson B, May D. Drip irrigation salinity management for row crops. Oakland: University of California; 2011.

[36] Slinger D, Tenison K. Salinity Glove Box Guide: NSW Murray & Murrumbidgee Catchments. New South Wales: NSW Department of Primary Industries; 2007.

[37] Grattan SR. Irrigation water salinity and crop production. Oakland: University of California; 2002.

[38] Edwards CA, Bohlen PJ. Biology and ecology of earthworms, London: Chapman & Hall; 1996.

[39] Podmore C. Irrigation salinity – causes and impacts. Primefact 2009;937(1) 1-4.

[40] FAO - Food and Agriculture Organization of the United Nations. Management of irrigation induced induced salt-affected soils. Rome: CISEAU, IPTRID, AGLL and FAO ; 1996. http://www.fao.org (accessed 07 August 2012).

[41] Prisco JT. Alguns aspectos da fisiologia do '' stress`` salino. Revista Brasileira de Botânica 1980, 3: 85-94.

[42] Prisco JT, Gomes-Filho E. Fisiologia e bioquímica do estresse salino em plantas. In : Gheyi HR, Dias NS, Lacerda FC. Manejo da Salinidade na Agricultura : Estudos Básicos e Aplicados (Eds). Expressão Gráfica e Editora : Fortaleza, INCTSal, 478p, 2010.

[43] Benzona N, Hensley D, Yogi J, Tavares J, Rauch F, Iwata R, Kellison M, Wong M, and Patti C. Salt and wind tolerance of landscape plants for Hawai'i. Cooperative Extension Service, Landscape 2009; 13 : 1-9.

[44] Silva EN, Ribeiro RV, Ferreira-Silva SL, Viégas RA, Silveira JAG. Comparative effects of salinity and water stress on photosynthesis, water relations and growth of *Jatropha curcas* plants. Journal of Arid Environments 2010; 74(10): 1-8.

[45] Sidari M, Mallamacci C, Muscolo A. Drought, salinity and heat differently affect seed germination of *Pinus pinea.* Journal of Forest Research 2008; 13:326–330.

[46] Wang W, Wang R, Yuan Y, Du N, Guo W. Effects of salt and water stress on plant biomass and photosynthetic characteristics of Tamarisk (*Tamarix chinensis* Lour.) seedlings. African Journal of Biotechnology 2011; 10: 17981-1789.

[47] Läuchli A, Grattan SR. Plant growth and development under salinity stress. In Jenkis MA, Hasegawa PM, Jain SM (Eds). Advances in Molecular Breeding Toward Drought and Salt Tolerant Crops. California: Springer; 2007. p. 1-32.

[48] Kayaa MD, Okçu G, Atak M, Çıkılı Y, Kolsarıcı O. Seed treatments to overcome salt and drought stress during germination in sunflower (*Helianthus annuus* L.). European Journal Agronomy 2006; 24: 291–295.

[49] Patanè C, Saita A, Sortino O. Comparative effects of salt and water stress on seed germination and early embryo growth in two cultivars of sweet sorghum. Journal Agronomy and Crop Science 2012, 1-8.

[50] Flowers TJ, Colmer TD. Salinity tolerance in halophytes. New Phytologist 2008; 179: 945–963.

[51] Albert R, Pfundner G, Hertenberger G, Kastenbauer T, Watzka M. The physiotype approach to understanding halophytes and xerophytes. In: Breckle S-W, Schweizer B, Arndt U, eds. Ergebnisse weltweiter ökologischer Forschung. Stuttgart, Germany: Verlag Günter Heimbach, 2000; p. 69–87.

[52] Breckle SW. Salinity, halophytes and salt affected natural ecosystems. In: Läuchli A, Lüttge U, eds. Salinity: environment–plants–molecules. Dordrecht, the Netherlands: Kluwer Academic Publishers; 2002. p. 53–77.

[53] Grigore MN, Villanueva M, Boscaiu M, Vicente O. Do Halophytes Really Require Salts for Their Growth and Development? An Experimental Approach. Notulae Scientia Biologicae 2012; 4(2):23-29.

[54] Vicente O, Boscaiu M, Naranjo MA, Estrelles E, Bellés JM, Soriano P. Responses to salt stress in the halophyte *Plantago crassifolia* (Plantaginaceae). Journal of Arid Environments, 2004; 58:463-481.

[55] Li XY, Zhang XM, He XY, Zeng FJ, Thomas FM, Foetzki A. Drought stress and irrigation effects on water relations of *Tamarix ramosissima* in the Qira Oasis. Acta Phytoecologica Sinica Chinese, 2004; 28(5): 644-650.

[56] Souza ER, Freire MBGS, Cunha KPV, Nascimento CWA, Ruiz HA, Teixeira MA. Biomass, anatomical changes and osmotic potential in *Atriplex nummularia* Lindl. cultivated in sodic saline soil under water stress. Environmental and Experimental Botany, 2012; 82: 20– 27.

[57] Silveira JAG, Araújo SAM, Lima JPMS, Viégas RA. Roots and leaves display contrasting osmotic adjustment mechanisms in response to NaCl-salinity in *Atriplex nummularia*. Environmental and Experimental Botany, 2009; 66: 1–8.

[58] Mamdouh M, Nemat A, Abdel-Hamid A, Mamdouh MS, Abu-Alnaga AZ, Nada RM, Reham M. Physiological aspects of tolerance in *Atriplex halimus* L. to NaCl and drought. Acta Physiologia Plantarum 2011; 33:547–557.

[59] Brown CE, Pezeshki SR, DeLauneb RD. The effects of salinity and soil drying on nutrient uptake and growth of *Spartina alterniflora* in a simulated tidal system. Environmental and Experimental Botany 2006; 58: 140–148.

[60] Da Matta F. Ecophysiology of tropical tree crops: an introduction. Brazilian Journal of Plant Physiology 2007; 19: 239-244.

[61] Burghardta M, Burghardta A, Gall J, Rosenbergerb C, Riedererb M. Ecophysiological adaptations of water relations of *Teucrium chamaedrys* L. to the hot and dry climate of xeric limestone sites in Franconia (Southern Germany). Flora 2008; 203: 3–13.

[62] Díaz-López, L.; Gimeno, V.; Lidón, V.; Simón, I.; Martínez, V.; García-Sánchez, F. The tolerance of *Jatropha curcas* seedling to NaCl: An ecophysiological analyses. Plant Physiology and Biochemistry 2012; 54: 34-24.

[63] Azevedo-Neto AD, Prisco JT, Enéas-Filho J, Lacerda CF, Silva JV, Costa PHA. Effects of salt stress on plant growth, stomatal reponse and solute accumulation of different maize genotypes. Brazilian Journal of Plant Physiology 2004; 16(1): 31-38.

[64] Esteves BS, Suzuki MS. Salinity effects on plants. Oecologia Brasilica 2008; 12 (4): 662-679.

[65] Bakke IA, Freire ALO, Bakke OA, Andrade AP, Bruno ALO. Water and sodium chloride effects on *Mimosa tenuiflora* (WILLD.) poiret seed germination. Caatinga 2006; 19(3): 261-267.

[66] Gondim FA, Gomes-Filho E, Lacerda CF, Prisco JT, Azevedo Neto AD, Marques EC. Pretreatment with H_2O_2 in maize seeds: effects on germination and seedling acclimation to salt stress. Brazilian Journal of Plant Physiology 2010; 22(2): 103-112.

[67] Silva EN, Silveira JAG, Rodrigues CRF, Lima CS, Viégas RA. Contribuição de solutos orgânicos e inorgânicos no ajustamento osmótico de pinhão-manso submetido à salinidade. Pesquia agropecuária brasileira 2009; 44(5): 437-445.

[68] Freitas VS, Alencar NLM, Lacerda CF, Prisco JT, Gomes-Filho E. Changes in physiological and biochemical indicators associated with salt tolerance in cotton, sorghum and cowpea. African Journal of Biochemistry Research 2011; 5(8): 264-271.

[69] Praxedes SC, Lacerda CF, DaMatta FM, Prisco JT, Gomes-Filho E. Salt tolerance is associated with differences in accumulation, biomass allocation and photosynthesis in cowpea cultivars. Journal of Agronomy and Crop Sceince 2010; 196: 193-204.

[70] Costa PHA, Silva JV, Bezerra MA, Enéas-Filho J, Prisco JT, Gomes-Filho E. Growth and organic and inorganic solute contents in NaCl-stressed cultivars of *Vigna unguiculata*. Revista Brasileira de Botânica 2003; 26 (3): 289-297.

[71] Marques EC, Freitas VS, Bezerra MA, Prisco JT, Gomes-Filho E. Effects of salt stress on germination, emergence and establishment of dwarf-cashew seedling. Revista Ciência Agronômica 2010; 42 (4): 993-999.

[72] Bezerra MA, Lacerda CF, Gomes-Filho E, Abreu CEB, Prisco JT. Physiology of cashew plants grown under adverse conditions. Brazilian Journal of Plant Physiology 2007; 19 (4): 449-461.

[73] Ferreira-Silva SL, Silveira JAG, Voigt EL, Soares LSP, Viégas RA. Changes in physiological indicators associated with salt tolerance in two contrasting cashew rootstocks. Brazilian Journal of Plant Physiology 2008; 20 (1): 51-59.

[74] Abreu CEB, Bezerra MA, Enéas-Filho J, Prisco JT, Gomes-Filho E. Physiological and biochemical changes occurring in dwarf-cashew seedlings subjected to salt stress. Brazilian Journal of Plant Physiology 2008; 20 (2): 105-118.

[75] Alvarez-Pizarro JC, Alencar NLM, Prisco JT, Gomes-Filho E. Salt-induced changes on H^+-ATPase activity, sterol and phospholipid content and lipid peroxidation of root plasma membrane from dwarf- cashew (*Anacardium occidentale* L.) seedlings. Plant Growth Regulation 2009; 59 (2): 125-135.

[76] Lima MA, Bezerra MA, Gomes-Filho E, Pinto CM, Enéas-Filho J. Gas exchange in leaves of sun and shade of cashew in different water regimes. Revista Ciência Agronômica 2010; 41 (4): 654-663.

[77] Amorim AV, Gomes-Filho E, Bezerra MA, Prisco JT, Lacerda CF. Physiologic responses of precocious dwarf cashew at different levels of salinity. Revista Ciência Agronômica 2010; 41(1): 113-121.

[78] Dombroski LD, Praxedes SC, Freitas RMO, Pontes FM. Water relations of Caatinga trees in the dry season. South African Journal of Botany 2011; 77: 430–434.

[79] Santos MG, Ribeiro RV, Machado EC, Pimentel C. Photosynthetic parameters and leaf water potential of five common bean genotypes under mild water deficit. Biologia Plantarum 2009, 53(2): 229-236.

[80] Trovão DMB, Fernandes PD, Andrade LA, Dantas-Neto J. Seazonal variations of physiological aspects of Caatinga species. Revista Brasileira de Engenharia Agrícola Ambiental 2007, 11(3): 307–311.

[81] Souza RP, Machado EC, Silva JAB, Lagôa AMMA, Silveira JAG. Photosynthetic gas exchange, chlorophyll fluorescence and some associated metabolic changes in cowpea (*Vigna unguiculata*) during water stress and recovery. Environmental and Experimental Botany 2004, 51: 45–56.

[82] Santos MG, Ribeiro RV, Machado EC, Pimentel C. Photosynthetic parameters and leaf water potential of five common bean genotypes under mild water deficit. Biologia Plantarum 2009, 53 (2): 229-236.

[83] Chaves MM, Maroco JP, Pereira JS. Understanding plant responses to drought-from genes to whole plant. Functional Plant Biology 2003, 30: 239-264.

Chapter 5

Silicon: A Benefic Element to Improve Tolerance in Plants Exposed to Water Deficiency

Allan Klynger da Silva Lobato,
Elaine Maria Silva Guedes,
Douglas José Marques and
Cândido Ferreira de Oliveira Neto

Additional information is available at the end of the chapter

http://dx.doi.org/10.5772/53765

1. Introduction

The silicon (Si) is an abundant element in terrestrial superficie [1], however its availability to plants is normally low [2]. According to Matichenkov & Calvert [3], the chemically active Si forms in soil are represented by soluble monosilicic acid ($Si(OH)_4$) that is soluble and weakly adsorbed, as well as acid polisilicic, which are compound organosilicates.

Si is considered an benefic element to higher plants [4], being that the absorption process must be active or passive [5], and deposition in cell walls of several organs such as leaf and stem can promote beneficial effects [6], and for this reason has been frequently linked to physiological, morphological, nutritional, and molecular aspects in plants [7-10].

In plants this nutrient is assimilated mainly by roots, and capacity to accumulate in tissues is variable [11], being several monocots such as *Oryza sativa* and *Triticum aestivum* considered silicon accumulator, with absorption active by root system, and it present leaf levels normally higher that 10.0 g kg^{-1} of Si [12]. In other hand, many dicots like as *Phaseolus vulgaris* and *Glycine max* are characterized as not accumulator of silicon, and its presents passive absorption, with leaf tenors minors that 5.0 g kg^{-1} of Si [13].

In tissues, about of 99% of silicon is found in silic form and less than 1% is colloidal or ionic form, which is the soluble form [14]. Therefore, the storage sites of silicon in plants normally are responsible to improve leaf and plant architectures and also others metabolic processes like as gas exchanges [15], photosynthetic pigments [16], and antioxidant system [17], in which it results in better performance linked to growth, development, and yield parameters [18] (Figure 1).

Figure 1. Effects of Si-treatment on growth of cvs. Hinohikari, Oochikara and lsi1 mutant. Aspects of wild type rice cvs. Hinohikari (a, d), Oochikara (b, e) and lsi1 mutant (c, f) were observed after control nutrient treatment of rice seedlings (a, b, c) and 14-day silicic acid treatment of rice seedlings (d, e, f) [7].

Drought is one of the key sources of abiotic stress, since it induces smaller growth and development rates, flower aborting, and decreases crop yields during the vegetative, reproductive, and maturation stages [19]. Usually, drought consequences on plants are studied in controlled/artificial conditions through water restriction.

Vascular plants present several strategies to minimize the negative effects induced by water deficiency, being morphological modifications like increase in root size [20] and reduction in leaf area [21]. Other responses are frequently reported, such as reductions in CO_2 assimilation by leaf through stomatal closing, membrane damage and disturbed activity of various enzymes, especially those of CO_2 fixation and adenosine triphosphate (ATP) synthesis. Enhanced metabolite flux through the photorespiratory pathway increases the oxidative load on the tissues as both processes generate reactive oxygen species. Injury caused by reactive oxygen species to biological macromolecules under drought stress is among the major deterrents to growth.

The stress occasioned by lower water supply to plant is defined as water deficit, being responsible to active responses in plant such as over-expression of genes linked to osmotic adjustment [22], modifications in physiological parameters as water potential, stomatal closing [23], and decrease in photosynthesis rate [24], besides reduction in cell metabolism with negative consequences on growth and production.

Gas exchanges like as leaf water potential, stomatal conductance, and transpiration rate has been used mainly to explain mechanisms affected during drought [25]. When the water availability in soil decrease, normally the water potentials of soil and leaf also are reduced, and it will provoke as consequence turgescence loss in plant cells, mainly in leaf, causing stomatal closing and limiting the gas exchanges [26].

The water potential describes the water amount, in which the water is moved in direction to potentials more negatives, and it differences of potential reveal water flux in soil-plant-atmosphere system [27]. In general, the leaf water potential is depending to soil water potential, water flux in system, and transpiration rate [28].

Attenuation of negative effects induced by the silicon application has been frequently described in plants exposed to water deficiency, such as Hattori et al. [2] investigating *Sorghum bicolor* plants related interference on gas exchanges. Results obtained by Ahmad & Haddad [29] working *Triticum aestivum* plants revealed influence positive on antioxidant enzyme activities. In addition, Lobato et al. [30] studying *Capsicum annuum* plants reported maximization in proline synthesis.

2. Objective

Aims of this chapter is to define (i) silicon and water deficit, to explain (ii) on silicon sources, uptake system and transporters into plant, and detection form in tissue, and to present (iii) the silicon action on gas exchanges and photosynthetic pigments in plants exposed to water restriction.

3. Silicon sources

The silicon uptake using metal salts of silicic acid normally requires their hydrolysis prior to their uptake. In either case they would affect the ionic balance of the system. The proposed mechanism for the solubilization of silica by PNO or MNO is novel and probably involves polarization of surface silica layer through interaction with the oxygen of the pyridine N-oxides. In the solubilization, PNO and MNO are regenerated, as evidenced by the fact that clear water containing freshly prepared PNO/MNO-silica complexes slowly deposits granular silica [31-32].

Based in overview related, a research was conduced by Ranganathan et al. [33] with *Oryza sativa* plants exposed to pyridine N-oxide (PNO), 4-morpholino pyridine N-oxide (MNO), and sodium metasilicate (SMS) aiming to test silicon sources on their ability to enhance the plant silicon content.

The suggestion that enhanced silica deposition is linked to the ability of the rice plant to combat abiotic and biotic stresses is further supported by SEM and EDX analysis of silicon distribution in the rice leaves. In the absence of added silicon, the leaves exhibited a scattered profile of silicon distribution (Figure 2). The leaves treated with MNO, PNO or SMS showed enhanced silicon content and localization of silicon bodies in leaf bundle sheath cells, particularly in the primary and secondary cell wall.

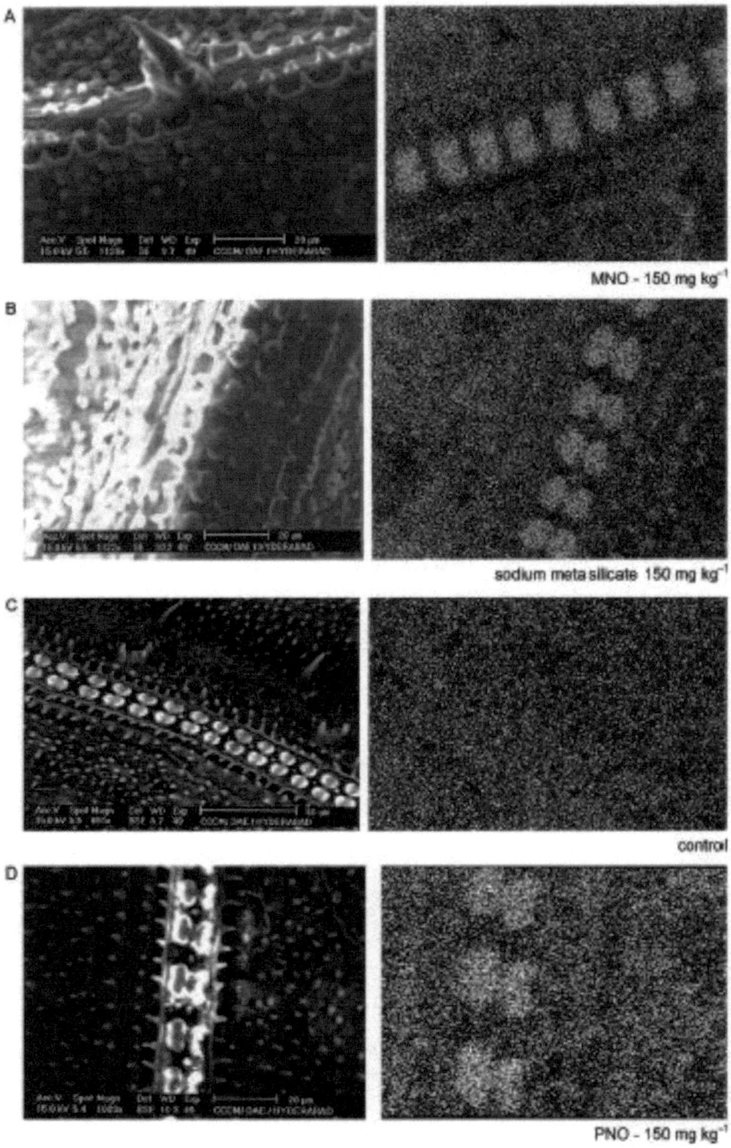

Figure 2. Scanning electron micrograph of silicon mapping (right) and its corresponding bundle sheath cells (left). Application of A - 150 mg kg^{-1} MNO, B - 150 mg kg^{-1} sodium silicate, C - control, and D - 150 mg kg^{-1} PNO [33].

Electron microscopy and in situ X-ray analysis of rice leaves reflect the differences in silicon distribution and cell wall structure between silicon treated and untreated plants [34-36]. The SEM pictures show that PNO and MNO enhanced the silica deposition on the leaves of rice plants concomitant with the localization of silicon bodies in leaf bundle sheath cells and in the primary and secondary cell walls [33].

4. Uptake system and transporters linked to silicon

For decades, rice has been known as the most effective silicon-accumulating species, although the mechanisms involved in high silicon uptake are poorly understood. One of the reasons is, unlike other minerals, the genotypic difference in silicon concentration of rice is too small to be utilized for comparative study on silicon uptake by rice roots [8]. With finality to resolve this problem, a research conduced by Ma et al. [8] working low-silicon mutant (*lsi1*) and wild-type of *Oryza sativa* plants revealed that there are least two transporters involved in silicon transport from nutritive solution to the xylem (Figure 3 and 4).

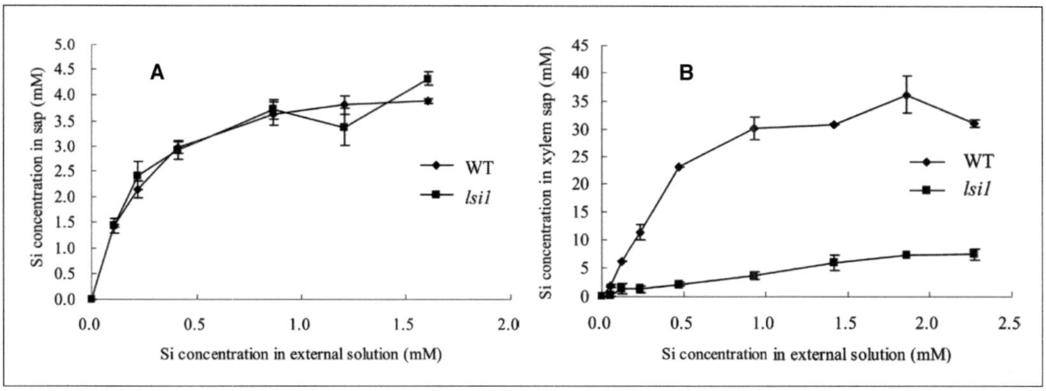

Figure 3. Concentration of silicon in the symplastic solution of *Oryza sativa* root tips (A), and concentration of silicon in the xylem sap (B) of rice cultured in silicon solution at various concentrations. Seedlings (26 d old) of wildtype (WT) and mutant (*lsi1*) rice were cultured in half-strength Kimura nutrient solution containing various concentrations of silicon. The stem was severed after 8 h, and the xylem sap was collected for 30 min. Values are means 6 SD of three replicates [8].

A kinetic study showed that the silicon concentration in the root-cell symplast increased with increasing silicon concentration in external solution but saturated at a higher silicon concentration in both lines (Figure 3 A). Again, there was no significant difference in the silicon concentration of symplastic solution between the wild type and the mutant. These results suggest that silicon transport from the external solution to the root cortical cells is mediated by a type of transporter and that the transporter of the mutant is identical to that of the wild type [8].

Other kinetic study on xylem loading of silicon was then conducted in the wild-type and mutant rice. In contrast to the silicon concentration in the root cortical cell symplast, the silicon concentration in the xylem sap was much higher in the wild type than in the mutant (Figure 3 B). In the mutant, the silicon concentration in sap increased gradually with increasing silicon concentration in the external solution without saturation. In the wild-type rice, the silicon concentration in the xylem sap also increased with increasing silicon concentration in the external solution (Figure 3 A), but it was saturated at a higher concentration [8].

The silicon concentration in the xylem sap of the wild type was higher than 30 mM at 0.9 mM silicon supply (Figure 3 B). This concentration was much higher than that in root-cell symplast (Figure 3 A), suggesting that silicon is transported from the root cells to the xylem also against a concentration gradient. The curve of Figure 2 B also suggests that the release of silicon into the xylem is mediated by a type of transporter in the wild type [8].

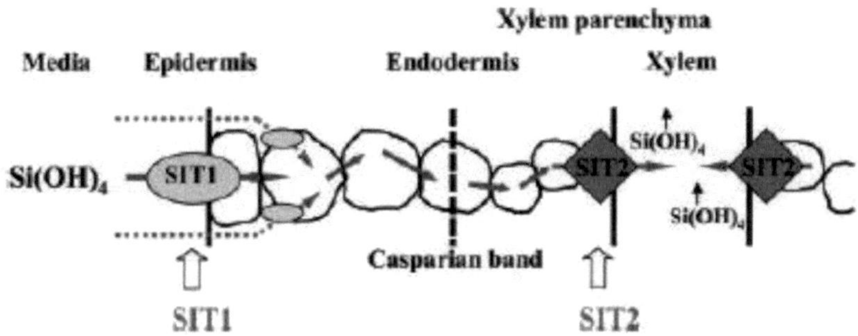

Figure 4. Schematic representation of the silicon uptake system in rice roots. SIT1, Silicon transporter from external solution to cortical cells. SIT2, Silicon transporter for xylem loading [8].

Above results suggest that at least two transporters are involved in the silicon uptake by rice roots (Figure 4). One is located on the plasma membrane of root cortical cells (SIT1, silicon transporter 1), which transport silicon from external solution to the root cortical cells. The other is located on the plasma membrane of xylem parenchyma cells (SIT2, silicon transporter 2), which is responsible for releasing silicon into the xylem. These transporters may have different affinities for silicic acid (Figure 3). Our results also clearly showed that the mutant is defective in xylem loading of silicon rather than transport of silicon from the external solution to the root cell [8].

5. Detection technique

Despite the abundance of studies of the effects of Si fertilizers and electron-microscopic observations of silica depositions within plants[13, 37-39], few of the morphological analysis of silica deposition in rice tissues have used histochemical staining and conventional microscopic techniques available to field researchers.

Although there are a few reports using X-ray scanning analytical microscopy, which described silica deposition detected in a dicot plant such as *Arabidopsis halleri* [40-41], X-ray scanning microscopy have been broadly applied for analysis of silica bodies and other elements in rice tissues [42-45].

Therefore, study carried out by Isa et al. [7] optimized a technique for silica body specific staining to visualize silica deposits in rice tissue by bright field microscopy and a technique for determining the in situ content of elements, including silica, by X-ray scanning microscopy.

The samples were fixed in FAA solution, and fixed samples were transferred and incubated in accordance with the methods of Kaufman et al. [46] and Morikawa & Saigusa [47], with minor modifications. Benzene-equilibrated samples were stained in 0.1% crystal violet lactone solution (in benzene) to visualize the silica bodies.

Silica bodies were observed in the motor cells of the leaf blades of cv. Hinohikari at the maximum tillering stage (Figure 5 A), and silica opal was also observed in the blades (Figure 5 B). In enlarged images of silica cells located along the vascular bundles of cv. Hinohikari treated with silicic acid for 14 days, accumulation of silica was observed as clear X shapes (Figure 5 C) [7].

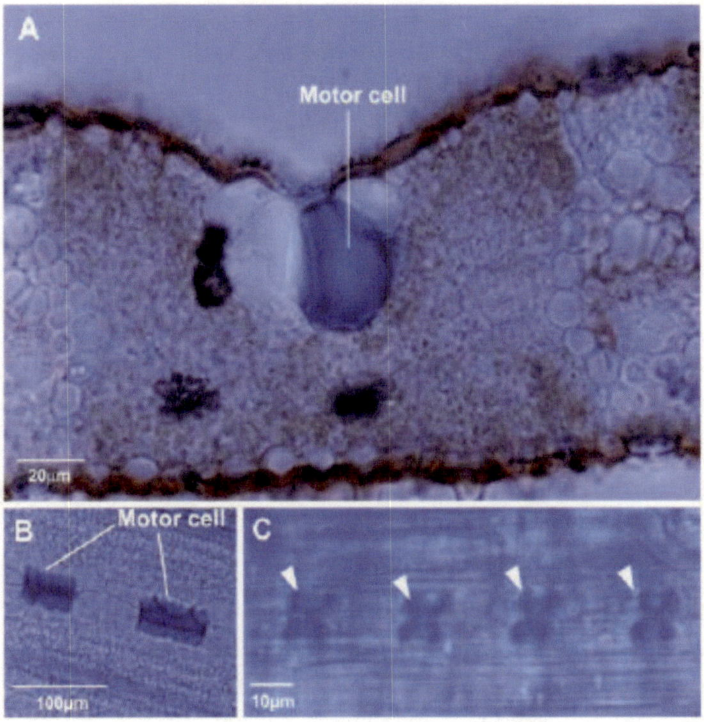

Figure 5. Crystal violet lactone staining of motor cell and silica body in leaf blade of cv. Hinohikari. A, motor cell and silica bodies in cross-section of leaf blade; B, rice opals in motor cells of leaf blade; C, a line of X-shaped silica bodies in leaf sheath. Rice opals in motor cells and silica bodies in epidermis cells were stained with crystal violet actone using leaf blade of cv. Hinohikari in paddy field at the maximum tillering stage. Motor cells and X-shaped silica bodies are indicated by lines and arrow heads, respectively [7].

Crystal violet lactone staining was an effective method of visualizing various shapes of silica opals and silica bodies deposited on the walls of the epidermal cells of leaf blades and stems, forming X-shaped silica cuticles along the vascular bundles (Figure 5 C) [7].

Successful staining reactions for observing silica bodies in plant tissues have been developed; they involve the use of methyl red, silver amine chromate, and crystal violet lactone [46, 48]. In accordance with the reactivity of the silanol groups on the surfaces of the silica bodies, the crystal violet lactone dye stains the silica bodies exclusively, clearly allowing their shapes to be observed [7].

6. Attenuation of negative effects produced by silicon in physiological parameters of plants exposed to water limitation

Benefits of silicon actuation recently reported on physiological parameters such as transpiration [10], stomatal conductance [49], and photosynthesis [50] were reported in several specie. In addition, pepper crops, more specifically *Capsicum annuum* exercises strong influency on Brazilian market, and there is necessacity to investigate silicon action on this crop. A study aiming to respond these questions was organized with five water and silicon combinations (control, deficit + 0.00, deficit + 0.25 deficit + 1.00, and deficit + 1.75 μM Si) applied to two cultivars (Ikeda and Vermelho Gigante) with a total of 10 treatments.

The water deficit promoted a decrease in leaf relative water content in two cultivars, and 0.25, 1.00, and 1.75 μM Si did not consistently increase this variable for Ikeda, although it was maintained at levels closer to the deficit. Leaf relative water content in Vermelho Gigante for all Si concentrations was slightly higher than the deficit + 0 μM Si (Figure 6 A).

Stomatal conductance was significantly reduced due to water deficit in both cultivars, compared with the control. Ikeda applied with 1.00 and 1.75 μM Si had higher stomatal conductance compared with deficit + 0 μM Si, while Vermelho Gigante had higher values at 0.25 and 1.75 μM Si (Figure 6 B).

The water deficit caused significant reduction in transpiration in Ikeda and Vermelho Gigante cultivars (Figure 6 C). Exogenous application of 0.25, 1.00, and 1.75 μM Si promoted attenuation of symptoms induced by water deficit. The treatments with added silicon were not statistically different.

The leaf relative water content of treatments under silicon application was maintained at levels closer to the control treatment, and this is linked to silicon action that was probably absorbed by plant, and deposited mainly in epidermal cell wall [51]. Additionally, the Si can contribute to higher resistance of xylem vessels [8], which are structures responsible by water transport into plant [52]. Therefore, plants with firmer xylem vessel walls can potentially avoid problems in these structures during drought or extreme heat, besides increasing water volume assimilated by plants [53]. Romero-Aranda et al. [54] investigating silicon effects on *Lycopersicon esculentum* plants under salt stress corroborate the data in this investigation.

http://dx.doi.org/10.5772/53765

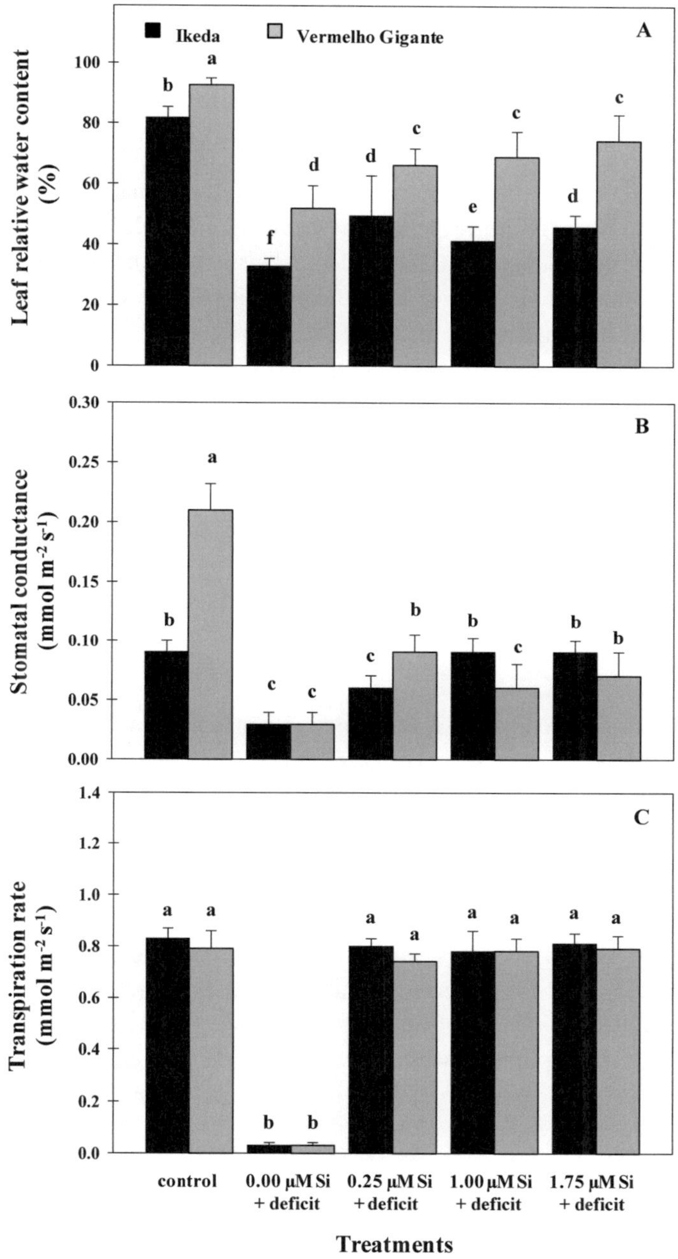

Figure 6. Leaf relative water content (A), stomatal conductance (B), and transpiration rate (C) in two pepper cultivars treated with silicon and exposed to water deficiency. Means followed by the same letter are not significantly different by the Scott-Knott test at 5% of probability. The bars represent the mean standard error [15].

The Si application attenuated the effects of water deficit, indicating intermediary levels in relation to stomatal conductance, and consequently, possible maintenance in gas exchange. A fall in this parameter will affect directly water relations, limiting the assimilation of carbon dioxide (CO_2) and water flux (H_2O) through stomata[55]. The stomatal mechanism will reduce the CO_2 assimilation, causing a reduction in photo-assimilate production and losses in yield [56]. Similar results were observed by Gong et al. [57] evaluating the silicon effects on *Triticum aestivum* plants under water deficit, with stomatal conductance being kept at intermediary levels in relation to control plants.

Silicon promoted attenuation in symptoms linked to water deficiency in transpiration rate, because during plant absorption in form of monosilicic acid (H_4SiO_4) [58], silicon accumulates in the leaf, forming a layer double of silicon. This accumulation promotes a reduction in transpiration and decrease water loss by the plant [59-60], but still contributing to maintenance of adequate transpiration rate. In addition, the transpiration process in plants is carried out by the stomatal present in leaf and cuticle normally in stem [61]. Agarie et al. [62] found improvement linked to transpiration rates in *Oryza sativa* plants cultivated with Si. Similar results on maintenance of transpiration were reported previously by Lobato et al. (2009a) studying the protective action of silicon in *Capsicum annuum* under water deficit.

7. Silicon promotes increase in chlorophylls of plants exposed to water deficit

Study conducted by Locarno et al. [63] described increase in level of chlorophyll a, and consequent repercussion on amount of chlorophyll total. Ávila et al. [64] investigating interaction between silicon and nitrogen reported also increase in level of chlorophyll a in *Oryza sativa* plants. In other hand, the drought normally promotes significant decrease in photosynthetic pigments [65-66].

Therefore, there is limited information linked to silicon action on chlorophyll levels in plants submitted to water deficit. Based in previous results described, Silva et al. [16] carried out an experiment with *Lycopersicon esculentum* plants submitted to water deficiency and silicon, being five water and silicon combinations (control, deficit + 0.00, deficit + 0.25, deficit + 1.00, and deficit + 1.75 μM Si) applied to two cultivars (Super Marmande and Santa Cruz) totalizing 10 treatments. This study had aim of explain silicon action on chlorophylls.

Water deficit promoted reduction not significant in relation to level of chlorophyll a in control treatment for Super Marmande, while Santa Cruz presented significant reduction (Figure 7 A). Applications of silicon in treatments deficit + 0.25, deficit + 1.00, and deficit + 1.75 μM Si promoted an increase in levels of chlorophyll a to both cultivars.

Cultivars evaluated under water deficit presented a reduction in chlorophyll b (Figure 7 B), when compared to control plants. Silicon applications in concentrations of 0.25, 1.00, and 1.75 μM Si provoked oscillations in this parameter. In treatments exposed to silicon, the levels of chlorophyll b presented better performances in concentrations of 1.00 and 1.75 μM Si for Super Marmande and Santa Cruz cultivars, respectively, being these treatments statistically equals to control to both cultivars.

Water deficiency occasioned significant decrease in total chlorophyll evaluated in both cultivars (Figure 7 C). Silicon in concentrations of 0.25, 1.00, and 1.75 μM Si induced a progressive increase in total chlorophyll. The better result was obtained in treatment exposed to concentration of 1.75 μM Si, and this treatment is statistically equal to control.

The reduction in chlorophyll a for both cultivars is a consequence of water restriction, in which will induce probably the production of peroxidative enzymes, and these enzymes are associated to degradation of chlorophyll a in thylakoid membrane [67]. In cultivars exposed to water deficit, silicon application proportioned increase in levels of chlorophyll a, indicating synthesis of new pigments, and maintenance of chlorophyll a previously existing. Donega [68] also affirmed that silicon uses promotes improvement in plant architecture and increase in photosynthesis. In other hand, the deposition of silicon in cell wall increased your tissue resistance, and it will keep plants with better performance linked to leaf position and interception of light [69-70]. Results of this research are similar and corroborate study carried out by Lobato et al. (2009a) investigating *Capsicum annuum* plants under water deficit, as well as Ahmad & Haddad [29] working with *Triticum aestivum* plants under water deficiency and silicon utilization reported similar results.

Water deficiency occasioned decrease chlorophyll b due to water restriction in substrate. Chlorophyll b was positively affected by silicon application, and this fact is associated to Si accumulation in epidermal cells localized in shoot, in which it will promote indirect protector effect on photosynthetic apparatus, and consequently decrease in damages provoked by water restriction in this parameter. The reduction in chlorophyll b in plants under water deficiency probably is resulting of disorders in chloroplast and modifications in relation proteins/lipids responsible by formation of pigment-protein complex [71-72]. Similar results on increase in chlorophyll b were observed by Locarno et al. [63] studying *Rosa* spp. plants under silicon application.

Results obtained in total chlorophyll indicate that plants under water deficit and treated with silicon presented increase in this parameter, and a relationship observed (data not shows) between leaf water potential and total chlorophyll can explain this fact, because adequate water amount in leaf tissue works probably maintaining stability of chloroplasts and consequently the functions carried out by the chlorophylls like as absorption and transport of energy.

In other hand, plants exposed to water deficit presented decrease, being probably linked to decrease in nitrogen absorption, in which is an essential element in formation of chlorophyll molecules. Plants under nitrogen application presented increase in chlorophyll amounts [64], as well as silicon promotes modifications on nitrogen metabolism [73]. Water works as vehicle responsible by nitrogen conduction and other nutrients during absorption through root system [74]. In other hands, during situations of water deficiency can there lower water assimilation, producing a negative interference linked to nitrogen assimilation. Therefore, this fact will generate minor amount of δ-aminolevulinate, which is chlorophyll precursor [75]. Tranaviciene et al. [76] investigating *Triticum aestivum* plants under different nitrogen levels observed that level of chlorophyll increased in consequence of increase in amount of nitrogen applied.

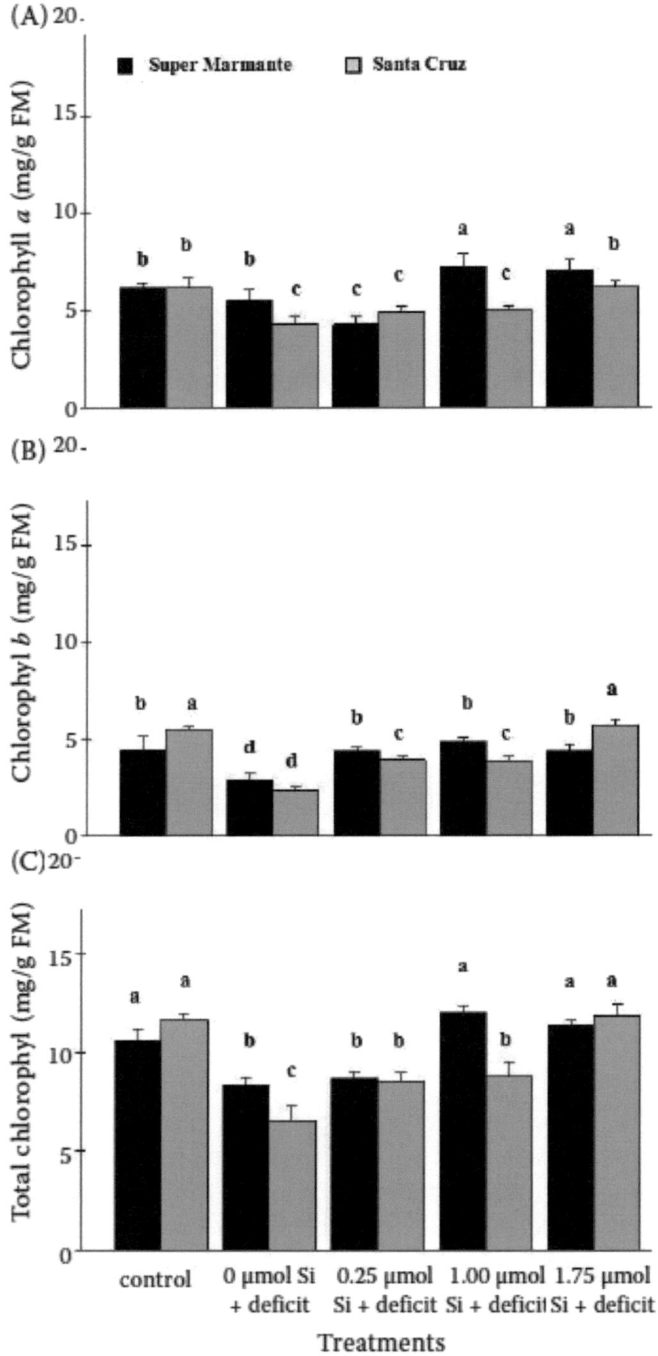

Figure 7. Chlorophyll a (A), chlorophyll b (B), and total chlorophyll (C) in two tomato cultivars treated with silicon and exposed to water deficiency. Means followed by the same letter are not significantly different by the Scott-Knott test at 5% of probability. The bars represent the mean standard error [16].

8. Final considerations

This chapter was structured with novel informations that can represent an important source to students, teachers, researchers, scientists and farmers on silicon action linked to attenuation of water deficit in higher plants. It revealed results and concepts on water deficiency and your consequences on plants, as well as silicon utilization with finality to improve tolerance during inadequate water supply. In addition, it presented a simple and efficient technique to carry detection of element beneficial in tissue. Based in recent results, was also demonstrated as silicon is assimilated, transported, and accumulated in several plant organs. Besides to prove positive interference on gas exchanges and photosynthetic pigments in plants exposed to water restriction.

Acknowledgements

This chapter had financial support from Conselho Nacional de Pesquisa (CNPq/Brazil) and Coordenação de Aperfeiçoamento de Pessoal de Nível Superior (CAPES/Brazil) for Lobato AKS.

Author details

Allan Klynger da Silva Lobato[1], Elaine Maria Silva Guedes[1], Douglas José Marques[2] and Cândido Ferreira de Oliveira Neto[1]

1 Núcleo de Pesquisa Vegetal Básica e Aplicada, Universidade Federal Rural da Amazônia, Paragominas, Brazil

2 Departamento de Ciência do Solo, Universidade Federal de Lavras, Lavras, Brazil

References

[1] Pereira HS, Korndörfer GH, Moura WF, Corrêa GF. Extractors of available silicon in slags and fertilizers. Revista Brasileira de Ciência do Solo 2003;27 265-274.

[2] Hattori T, Inanaga S, Araki H, An P, Morita S, Luxová M, Lux A. Application of silicon enhanced drought tolerance in *Sorghum bicolor*. Physiologia Plantarum 2005;123 459-466.

[3] Matichenkov VV, Calvert DV. Silicon as a beneficial element for sugarcane. Journal American Society of Sugarcane Technologists 2002;22 21-30.

[4] Epstein E, Bloom AJ. Mineral nutrition of plants: principles and perspectives. Sinauer Associates, Sunderland, 2004.

[5] Ma JF, Yamaji N, Mitani N, Tamai K, Konishi S, Fujiwara T, Katsuhara M, Yano M. An efflux transporter of silicon in rice. Nature 2007;448 209-212.

[6] Cunha KPV, Nascimento CWA, Accioly AMA, Silva AJ. Cadmium and zinc availability, accumulation and toxicity in maize grown in a contaminated soil. Revista Brasileira de Ciência do Solo 2008;3 1319-1328.

[7] Isa M, Bai S, Yokoyama T, Ma JF, Ishibashi Y, Yuasa T, Iwaya-Inoue M. Silicon enhances growth independent of silica deposition in a low-silica rice mutant, *lsi1*. Plant Soil 2010;331 361-375.

[8] Ma JF, Mitani N, Nagao S, Konishi S, Tamai K, Iwashita T, Yano M. Characterization of the silicon uptake system and molecular mapping of the silicon transporter gene in rice. Plant Physiology 2004;136 3284-3289.

[9] Ma JF, Yamaji N. Silicon uptake and accumulation in higher plants. Trends in Plant Science 2006;11 342-397.

[10] Lobato AKS, Coimbra GK, Neto MAM, Costa RCL, Santos Filho BG, Oliveira Neto CF, Luz LM, Barreto AGT, Pereira BWF, Alves GAR, Monteiro BS, Marochio CA. Protective action of silicon on relations and photosynthetic pigments in pepper plants induced to water deficit. Research Journal of Biological Sciences 2009a;4 617-623.

[11] Chiba Y, Mitani N, Yamaji N, Ma JF. *HvLsi 1* is a silicon influx transporter in barley. Plant Journal 2009;57 810-818.

[12] Oliveira LA. Silicon in bean and rice plants: absorption, transport, redistribution, and tolerance to cadmium. [PhD. Thesis] Piracicaba, Escola Superior de Agricultura Luiz Queiroz: p. 1-157, 2009.

[13] Takahashi E, Ma JF, Miyake Y. The possibility of silicon as an essential element for higher plants. Comments on Agricultural and Food Chemistry 1990;2 99–122.

[14] Ma JF, Miyake Y, Takahashi E. Silicon as a beneficial element for crop plants. In: Datnoff LE, Snyder GH, Korndörfer GH (ed.). Silicon in agriculture. Amsterdan: Elsevier, 2001. p. 17-39.

[15] Pereira TS, Lobato AKS, Tan DKY, Costa DV, Uchôa EB, Ferreira RN, Pereira ES, Ávila FW, Marques DJ, Guedes EMS. Positive interference of silicon on water relations, nitrogen metabolism, and osmotic adjustment in two pepper (*Capsicum annuum*) cultivars under water deficit. Australian Journal of Crop Science 2012;paper in press.

[16] Silva ON, Lobato AKS, Martins Filho AP, Lemos RP, Pinho JM, Medeiros MBCL, Cardoso MS, Ávila FW, Costa RCL, Oliveira Neto CF, Santos Filho BG, Andrade IP. Silicon contributes to increase chlorophyll and this response is modulated by leaf wa-

ter potential in two tomato cultivars exposed to water deficiency. Plant, Soil, and Environment 2012;paper is press.

[17] Li P, Song A, Li Z, Fan F, Liang Y. Silicon ameliorates manganese toxicity by regulating manganese transport and antioxidant reactions in rice (*Oryza sativa* L.). Plant Soil 2012;354 407-419.

[18] Nolla RJF, Korndörfer GH, Silva TRB. Effect of silicon on drought tolerance of upland rice. Journal of Food, Agriculture & Environment 2012;10 269-272.

[19] Showemimo FA, Olarewaju JD. Drought tolerance indices in sweet pepper (*Capsicum annuum* L.). International Journal of Plant Breeding and Genetics 2007;1 29-33.

[20] Lobato AKS, Costa RCL, Oliveira Neto CF, Santos Filho, BG, Cruz FJR, Freitas JMN, Cordeiro FC. Morphological changes in soybean under progressive water stress. International Journal of Botany 2008;4 231-235.

[21] Anyia AO, Herzog H. Water-use efficiency, leaf area and leaf gas exchange of cowpeas under mid-season drought. European Journal of Agronomy 2004;20 327-339.

[22] Rodrigues FA, Laia ML, Zingaretti SM. Analysis of gene expression profiles under water stress in tolerant and sensitive sugarcane plants. Plant Science 2009;176 286-302.

[23] Santos MG, Ribeiro RV, Oliveira RF, Machado EC, Pimentel C. The role of inorganic phosphate on photosynthesis recovery of common bean after a mild drought deficit. Plant Science 2006;170 659-664.

[24] Hong-Bo S, Chu LY, Jaleel CA, Zhao CX. Water-deficit stress-induced anatomical changes in higler plants. Comptes Rendus Biologies 2008;331 215-225.

[25] Gonçalves ER, Ferreira VM, Silva JV, Endres L, Barbosa TP, Duarte WG. Gas exchange and chlorophyll a fluorescence of sugarcane varieties submitted to water stress. Revista Brasileira de Engenharia Agrícola e Ambiental 2010;14 378-386.

[26] Gubiani P. Time to occurrence of penetration resistance to bean estritiva in soil with different states of compression. [MSc. Dissertation] Santa Maria, Universidade Federal de Santa Maria. p. 1-108, 2008.

[27] Martins J. Morphophysiological modifications in maize plants subjected to water deficit. [MSc. Dissertation] Santa Maria, Universidade Federal de Santa Maria. p. 1-102, 2010.

[28] Yu GR, Zhuang J, Nakayama K, Jin Y. Root water uptake and profile soil water as affected by vertical root distribution. Plant Ecology 2007;189 15-30.

[29] Ahmad ST, Haddad R. Study of silicon effects on antioxidant enzyme activities and osmotic adjustment of wheat under drought stress. Czech Journal of Genetics and Plant Breeding 2011;47 17–27.

[30] Lobato AKS, Luz LM, Costa RCL, Santos Filho BG, Meirelles ACS, Oliveira Neto CF, Laughinghouse HD, Neto MAM, Alves GAR, Lopes MJS, Neves HKB. Silicon exer-

cises influence on nitrogen components in pepper subjected to water Deficit? Research Journal of Biological Sciences 2009b;4 1048-1055.

[31] Chandrasekher Rao C. Molecular Organization and Self Assembly. - Ph.D. Thesis. Osmania University, Hyderabad 2002.

[32] Ranganathan S, Chandrasekhara Rao C, Suvarchala V, Rajesh YBRD, Jagdesh B. Solubilization of silica: synthesis, characterization and study of penta-coordinated pyridine N-oxide silicon complexes. - Journal of Chemical Sciences 2004;116 169-174.

[33] Ranganathan S, Suvarchala V, Rajesh YBRD, Srinivasa Prasad M, Padmakumari AP, Voleti SR. Effects of silicon sources on its deposition, chlorophyll content, and disease and pest resistance in rice. Biologia Plantarum 2006;50 713-716.

[34] Balasta MLFC, Perez CM, Jullano BO, Villareal CP, Lott JNA, Roxas DB. Effects of silica level on some properties of *Oryza sativa* straw and hull. Canadian Journal of Botany 1989;67 2356-2363, 1989.

[35] Kim SG, Kim KW, Park EW, Choi D. Silicon-induced cell wall fortification of rice leaves: a possible cellular mechanism of enhanced host resistance to blast. Phytopathology 2002;92 1095-1103.

[36] Rodrigues FA, Benhamou N, Datnoff LE, Jones JB, Belanger RR. Ultrastructural and cytochemical aspects of silicon-mediated rice blast resistance. Phytopathology 2003;93 535-546.

[37] Tsuda M, Morita M, Mkihara D, Hirai Y. The involvement of silicon deposition in salinity-induced white head in rice (*Oryza sativa* L.). Plant Prod Sci 2000;3 328–334.

[38] Agarie S, Agata W, Kaufman PB. Involvement of silicon in the senescence of rice leaves. Plant Production Science 1998a;1 104–105.

[39] Agarie S, Agata W, Uchida H, Kubota F, Kaufman PB. Function of silica bodies in the epidermal system of rice (*Oryza sativa* L): testing the window hypothesis. Journal of Experimental Botany 1998b;47 655–660.

[40] Zhao FJ, Lombi E, Breedon T (2001) Zinc hyperaccumulation and cellular distribution in *Arabidopsis halleri*. Plant Cell Environ 2001;23 507–514.

[41] Hokura A, Onuma R, Kitajima N, Terada Y, Saito H, Abe T, Yoshida S, Nakai I. 2-D X-ray fluorescence imaging of cadmium hyperaccumulating plants by using high-energy synchrotron radiation X-ray microbeam. Chemistry Letters 2006;35 1246–1250.

[42] Kaufman PB, Soni SL, LaCroix JD, Rosen JJ, Bigelow WC. Electron-probe microanalysis of silicon in the epidermis of rice (*Oryza sativa* L.). Planta 1972;104 10–17.

[43] Soni SL, Kaufman PB, Bigelow WC. Electron probe analysis of silicon and other elements in leaf epidermal cells of the Rice plant (*Oryza sativa* L). American Journal of Botany 1972; 59 38–42.

[44] Dikeman E, Bechtel DB, Pomeranz Y. Distribution of elements in the rice kernel determined by X-ray analysis and atomic absorption spectroscopy. Cereal Chemistry 1981;58 148–152.

[45] Takahashi N, Kato Y, Isogai A, Kurata K. Silica distribution on the husk epidermis at different parts of the panicle in rice (*Oryza sativa* L) determined by X-ray microanalysis. Plant Production Science 2006;9 168–171.

[46] Kaufman PB, Dayanandan P, Franklin CI, Takeoka Y. Structure and function of silica bodies in the epidermal system of grass shoots. Annals of Botany 1985;55 487–507.

[47] Morikawa CK, Saigusa M. Mineral composition and accumulation of silicon in tissues of blueberry (*Vaccinum corymbosus* cv. Bluecrop) cuttings. Plant Soil 2004;258 1–8.

[48] Dayanandan P, Kaufman PB, Franklin CI. Detection of silica in plants. American Journal of Botany 1983;70 1079–1084.

[49] Gunes A, Inal A, Bagci EG, Coban S, Pilbeam DJ. Silicon mediates changes to some physiological and enzymatic parameters symptomatic for oxidative stress in spinach (*Spinacia oleracea* L.) grown under B toxicity. Scientia Horticulturae 2007;113 113-119.

[50] Sacala E. Role of silicon in plant resistance to water stress. Journa of Elementology 2009;14 619-630.

[51] Savant NK, Datnoff LE, Snyder GH. Depletion of plant-available silicon in soils: a possible cause of declining rice yields. Communications in Soil Science and Plant Analysis 1997;28 1245-1252.

[52] McElrone AJ, Pockman WT, Martínez-Vilata J, Jackson RB. Variation in xylem structure and function in stems and roots of trees to 20 m depth. New Phytologist 2004;163 507-517.

[53] Sperry JS, Iiacke UG, Oren R, Comstock JP. Water deficits and hydraulic limits to leaf water supply. Plant, Cell & Environment 2002;25 251-263.

[54] Romero-Aranda MS, Jurado O, Cuartero J. Silicon alleviates the deleterious salt effect on tomato plant growth by improving plant water status. Journal of Plant Physiology 2006;163 847-855.

[55] McDermit DK. Sources of error in the estimation of stomatal conductance and transpiration from porometer data. HortScience 1990;25 1538-1548.

[56] Paiva AS, Fernandes EJ, Rodrigues TJD, Turco JEP. Stomatal conductance in leaves of bean plants submitted to different irrigation regimes. Engenharia Agrícola 2005;25 161-169.

[57] Gong H, Zhu X, Chen K, Wang S, Zhang C. Silicon alleviates oxidative damage of wheat plants in pots under drought. Plant Science 2005;169 313-321.

[58] Richmond KE, Sussaman M Got silicon? The non-essential beneficial plant nutrient. Current Opinion in Plant Biology 2003;6 268-272.

[59] Freitas LB, Coelho EM, Maia SCM, Silva TRB. Foliar fertilization with silicon in maize. Revista Ceres 2011;58 262-267.

[60] Datnoff LE, Snyder GH, Korndörfer GH. Silicon on Agriculture. Elsevier Science, Amsterdam. 2001.

[61] Kerbauy GB. Plant Physiology. Guanabara Koogan S.A., Rio de Janeiro. 2004.

[62] Agarie S, Hanaoka N, Ueno O, Miyazaki A, Kubota F, Agata W, Kaufman PB. Effects of silicon on tolerance to water deficit and heat stress in rice plants (*Oryza sativa L.*) monitored by electrolyte leakage. Plant Production Science 1998c;1 96-103.

[63] Locarno M, Fochi CG, Paiva PDO. Influence of silicate fertilization on chlorophylls of rose leaves. Ciência e Agrotecnologia 2011;35 287-290.

[64] Ávila FW, Baliza DP, Faquin V, Araujo J, Ramos SJ. Silicon-nitrogen interaction in rice cultivated under nutrient solution. Revista Ciencia Agronomica 2010;41 184-190.

[65] Lobato AKS, Costa RCL, Neto MAM, Oliveira Neto CF, Santos Filho BG, Alves GAR, Freitas JMN, Cruz FJR, Marochio CA, Coimbra GK. Responses of the photosynthetic pigments and carbon metabolism in *Vigna unguiculata* cultivars submitted to water deficit. Research Journal of Biological Sciences 2009c;4 593-598.

[66] Oliveira Neto CF, Lobato AKS, Gonçalves-Vidigal MC, Costa RCL, Santos Filho BG, Alves GAR, Maia WJMS, Cruz FJR, Neves HKB, Lopes MJS. Carbon compounds and chlorophyll contents in sorghum submitted to water deficit during three growth stages. Journal of Food, Agriculture & Environment 2009;7 588-593.

[67] Gandul-Rojas B, Roca M, Mínguez-Mosquera MI. Chlorophyll and carotenoid degradation mediated by thylakoid-associated peroxidative activity in olives (*Olea europaea*) cv. Hojiblanca. Journal of Plant Physiology 2004;161 499-507.

[68] Donegá MA. Ratio K: Ca and application of silicon in the nutrient solution for the hydroponic cultivation of coriander. [MSc. Dissertation] Piracicaba, Escola Superior de Agricultura Luiz Queiroz. p. 1-62, 2009.

[69] Fiori MP. Behavior of tomato cultivars as to use of steel slag in environment protected. [MSc. Dissertation] Marilia, Universidade de Marília. p. 1-54, 2006.

[70] Lana RMQ, Korndorfer GH, Zanão Júnior LA, Silva AF, Lana AMQ. Effect of calcium silicate on the productivity and silicon accumulation in the tomato plant. Bioscience Journal 2003;19:15-20.

[71] Parida A, Das AB, Das P. NaCl stress causes changes in photosynthetic pigments, proteins and other metabolic components in the leaves of a true mangrove, *Bruguiera parviflora*, in hydroponic cultures. Journal of Plant Biology 2002;45 28-36.

[72] Parida AK, Dagaonkar VS, Phalak MS, Auramgabadkar LP. Alterations in photosynthetic pigments, protein and osmotic components in cotton genotypes subjected to

short-term drought stress followed by recovery. Plant Biotechnology Reports 2007;1 37–48.

[73] Watanabe S, Fujiwara T, Yoneyama T, Hayashi H. Effects of silicon nutrition on metabolism and translocation of nutrients in rice plants. Developments in Plant and Soil Sciences 2002;92 174-175.

[74] Santos RF, Carlesso R. Water deficit and morphologic and physiologic behavior of the plants. Revista Brasileira de Engenharia Agrícola e Ambiental 1998;2 287-294.

[75] Siqueira SC, Moreira MA, Mosquim PR, José IC, Ferreira FA, Sediyama CS. Simulation of the transgenic soybean tolerant to glyphosate through explant cultivation. Planta Daninha 1999;17 95-107.

[76] Tranaviciene T, Urbonaviciute A, Samuoliene G, Duchovskis P, Vaguseviciene I, Sliesaravicius A. The effect of differential nitrogen fertilization on photosynthetic pigment and carbohydrate contents in the two winter wheat varieties. Agronomy Research 2008; 6 555-561.

Chapter 6

Water Stress in Small Ruminants

Lina Jaber, Mabelle Chedid and Shadi Hamadeh

Additional information is available at the end of the chapter

http://dx.doi.org/10.5772/53584

1. Introduction

Small ruminants are an integral part of farming systems in the marginal arid regions of the world. These areas are characterized by water scarcity and fluctuating precipitation; under the effect of global warming and unpredictable weather, rainfall is becoming even more irregular and water availability more limited. Along with water accessibility, feed and other resources will be markedly affected by climate change. Livestock that are able, in open range, to select high quality forage to maintain a relatively similar basal diet quality from season to season, will have their intake significantly reduced in extremely dry seasons when forage biomass and its quality are low [1]. Hence, selection of adapted animal breeds is very valuable for sustaining animal production under an increasingly challenging environment [2].

Breeds of ruminants which are well adapted to arid environments demonstrate a greater capability than non-desert breeds to endure the stressful environmental effects [3]. Although small ruminants in hot arid and semi-arid regions may survive up to one week with little or even no water, water deficiency is proved to affect animals' physiological homeostasis leading to loss of body weight, low reproductive rates and a decreased resistance to diseases [4]. In addition, under natural conditions, water scarcity often occurs at times of high environmental temperature and low feed quality and availability. Therefore, the effects of these three constraints are often confounded.

In this review, the effects of various degrees and forms of dehydration on small ruminants are presented. The findings are based on previous literature on the subject as well as the results of original research by the authors on water restriction in Awassi sheep. The major changes in physiological indicators and blood parameters are presented, in addition to the interaction of dehydration with physiological status. Finally, a brief overview of new approaches for water stress alleviation, through drugs, is exposed.

2. General characteristics of small ruminants in arid and semi-arid regions

Small ruminants in the Middle East and North Africa region are mostly reared under extensive and traditional pastoral farming systems, centuries-old, relying on natural pastures and mobility to secure water and feed year round. Giger-Reverdin and Gihad (1991) [5] reported that the water requirement for maintenance of goats in temperate climates is 107ml/kg $BW^{0.75}$; they also indicated that the water requirements under different ambient temperatures, based on previous work, range between 3.15 kg/kg DM (at 23°C) to 4.71 kg/kg DM (at 35°C). However, the effect of ambient temperature should be viewed in combination with the particular breed origin and adaptability, since wide differences in response to heat have been reported in different breeds [6].

Sheep breeds differ in their capacity to overcome water limitation; experiments show a variety of results: Yankasa sheep survived 5 days of water restriction but with several physiological changes [7, 8]. Jaber et al. (2004) [9] concluded that Awassi females can withstand more than one month of watering every 2 days without significant changes, while a regime of watering once every five days causes important physiological perturbations. The Australian Merino sheep survived 10 days without water [10], and the desert bighorn sheep withstood water deprivation up to 15 days [11; 12], while the Barki sheep in Egypt did not endure 3 days without drinking [11].

Similarly, variations in water deprivation tolerance are observed in goat breeds. Ahmed and El Kheir (2001) [13] report that desert goats raised under traditional systems may be watered only once every 3-6 days, when water is scarce. The Black Bedouin and the Barmer goats are another example of adapted breeds that can live on a once every four days watering regime [14; 15].

Review papers are numerous on this subject, indicating differences between adapted and non-adapted sheep and goat breeds in tolerating water deprivation and the general arid conditions [6, 16].

Indigenous small ruminants are able to thrive despite extreme temperatures and limited water through their behavioral adaptations in combination with both morphological and physiological adaptations [17].

2.1. Behavioral adaptations

Feeding behavior is affected by environmental constraints. Nocturnal feeding has been documented in bighorn sheep [18] in order to avoid high temperatures during the day. Similar behavior is also reported in goats [16]. The author also indicated that feeding frequency is modified in some adapted goats which resort to more frequent and shorter meals in order to reduce heat production associated with rumen fermentation. Langhans et al. (1991) [19] further observed that feed intake is less affected by water deprivation in adapted breeds such as pygmy goats as compared to non-adapted breeds. Furthermore, Lechner-doll et al. (1995) [20] observed that adapted goats can select high quality feed during the dry season while sheep showed less selectivity to high quality feed. Drinking behavior is also affected by water restriction whereby water deprived sheep and goats tend to drink large volumes of water in one bout upon watering. This capacity is more pronounced in goats than in sheep [5].

Timing of reproduction is another adaptive feature of sheep living in semi-arid areas: parturition is timed to a favorable period of the year to ensure offspring survival [21]. Seasonality in small ruminants is commonly observed. Reproductive cycles are thought to be regulated by environmental cues, most importantly photoperiod. Research has demonstrated that nutrition and the general body score are also important factors for normal cyclicity. Under-nutrition (below 40-60% of maintenance requirement) is reported to cause an immediate retardation in follicular growth [22]. In addition, prolonged under-nutrition induces a delay in estrous behavior that lasts for a shorter period, as compared to normally fed animals. Similarly, fasting was reported to cause major changes in the concentrations of reproductive hormones and ovulation rate [23, 24]. Since water deprivation is often accompanied by feed intake reduction, the same effects on ruminant reproductive cycles can be expected. Indigenous sheep living in the tropical and sub-tropical regions tend to breed throughout the year; however, their sexual activity may be limited, to a degree, during the summer season when the environmental temperature is elevated and feed is lacking [25]. In arid and semi-arid regions, where differences in daylight, as well as in food and water availability are well marked, the breeding season usually spans from June to November [26, 27]. Consequently, kidding and lambing mostly occur between February and April, when food and climate become more hospitable for the newborns.

2.2. Morphological adaptations

Ruminants are usually classified as grazers, browsers or intermediate feeders. Sheep are usually classified as grazers feeding mainly on grasses while goats are intermediate feeders which can use grasses as well as shrubs [28]. These observed preferences in feed selection have been linked to underlying morphological and physiological digestive differences. These include larger rumen and long feed passage time for grazers to allow them to digest their high fiber diet, while browsers have simple and smaller digestive system with profuse saliva production to effectively process their feed high in cell solubles [28]. However, other scientists have argued that the observed differences in feed selection are related more to body size than to actual differences in the digestive anatomy and physiology [29]. Other morphological differences are also noted in relation to the mouth anatomy with goats having a mobile upper lip, while sheep are characterized by a cleft upper lip, features that allow them to best use the available vegetation. In addition, goats have the capacity to assume a bipedal position thus they are capable of browsing higher vegetation that is beyond the reach of sheep [30]. This area of research relating morphology, diet selection and utilization and response to changes in the available vegetation is still in need of further exploration [29, 31] as it is affected by the climatic changes and drought spells that put pressure on the vegetation cover and the animals that feed on it.

Morphological characteristics such as body shape and size help reducing heat loads and minimizing water losses; it is noted that goat breeds of arid and semi-arid regions are relatively smaller than their European counterparts [6]. Smaller animals benefit from a relatively larger surface area which allows them to better dissipate heat to the environment. Fleece is another feature that plays a major role in controlling body temperature, serving as a thermal barrier that reduces the effects of the ambient temperatures through the formation of a milder microclimate within the fleece [32]. In addition, fleece and hair color of small ruminants play a role in reflecting solar radiation with light colors absorbing less heat that the darker ones thus leaving the underneath skin relatively cooler [33]. Therefore, thermostability could be maintained without directly resorting to evaporative cooling (panting) which leads to high water loss [34]. Location of body fat also affects heat dissipation rates: arid-adapted sheep exhibit highly localized fat storage [35], such as in the fat-tail, as opposed to high subcutaneous fat in non-adapted breeds; this again facilitates heat conductance to the periphery for dissipation [34]. Moreover, fat-tails are important energy reserves that help in buffering long-term dietary shortfalls to maintain survival and productivity [36, 37]. The Awassi sheep, as a representative breed adapted to arid regions, presents a medium body size, with a large fat-tail and carpet type fleece. The carpet-type wool of the Awassi allows convective heat loss from the skin to the environment [34]. Their large ears are another anatomical adaptation that is thought to help in convective heat loss [34]. Finally, as mentioned above, the localization of the fat stores in the fat-tail facilitates body heat dissipation and serves as energy reservoir for times of scarcity.

2.3. Physiological adaptations

Physiologically, ruminant breeds of arid regions show many adapted mechanisms to conserve water in times of heat and drought. Adapted breeds resort to reduction of urine volume and fecal moisture. The production of more concentrated urine is related to the length of Henlé loops located in the medulla of the kidney [38]. The thickness of the medulla is relative to kidney size, and is frequently used as an index of kidney concentrating ability [38, 39]. For instance, the desert bighorn sheep exhibits a medulla nearly twice thicker than that of other domestic sheep and thus produces highly concentrated urine of 3900 mOsm/liter H_2O [40, 41]. The Awassi sheep demonstrated a similar ability to highly concentrate urine (up to 3244 mOsm/Kg H_2O) under dehydration, and to drink large volumes upon rehydration without disrupting their homeostasis [42].

Urea renal retention is similarly increased under dehydration leading to increased urea concentration in the blood, on the other hand, urea recycling from the blood in to the gut is often observed under these conditions and is thought to contribute as a nitrogen source in times when the quality of the offered feed is low in protein [33]. The rumen is another organ that plays an important role in maintaining homeostasis under dehydration in adapted ruminants, particularly goats. Due to its relatively large volume, it acts as an important water reservoir providing most of the water lost during prolonged dehydration to maintain blood volume. It also allows the intake of large volumes of water upon rehydration which is temporarily sequestered in the rumen. Through efficient and well-coordinated mechanisms of saliva recycling and high water and Na^+ retention in the kidneys, slow rehydration is achieved without causing water toxicity and with minimal water losses. These processes are detailed in [15]. Small increases in body temperature are also observed during the hottest parts of the day, followed by body cooling at night through conduction and radiation. The capacity to tolerate this increase in temperature means that less water is needed for evaporative cooling [33].

3. Physiological changes in response to water stress

3.1. Effect on feed intake and body weight

Studies show a close relation between water intake and feed consumption [3, 43, 44]. Ruminant feeding behavior can be affected by the changes in osmolality of body fluids [45]. Feed intake causes hypovolemia and hyperosmolality due to the secretion of saliva and gastric juices. These mechanisms can urge ruminants, as well as other animals, to drink while eating, or alternatively not to eat when severely dehydrated [45]. Moreover, an adequate level of water intake is necessary for proper digestive functions [46]. On the other hand, Kay (1997) [33] states that drinking water is not needed for swallowing and moistening feed, since water can be circulated from the blood to maintain high salivation. However, water is needed to replace the inevitable water loss by excretion and evaporation. A possible explanation for the physiological mechanism behind the reduction in feed intake under water restriction, mainly through the reduction in meal size, may be attributed to the postprandial hyperosmolality of the ruminal fluid [19].

The co-occurrence of decrease in feed intake along with water restriction renders the differentiation between water versus feed shortage related effects difficult. Previous work has shown that Awassi sheep under 3-4 days intermittent watering regime reduced their voluntary feed intake to approximately 60% of the control [9, 47]. Similar rates are reported in [48] in different domestic ruminants subjected to dehydration, especially when combined with heat stress. The drop in feed intake under dehydration is also dependent on the type of feed that is available to the animals. Water restricted goats reduced their feed intake by 18.8% when offered legume hay compared to 21.21% when offered grass hay with lower crude protein content [49]. Therefore the negative effect of water restriction is more pronounced on low versus high quality forage [16]. This reduction in feed intake is partially compensated for by a slower feed movement and longer retention time in the digestive tract [46, 50, 51]. This is thought to lead to an increase in digestibility and nutrient utilization, as longer time is available for the microflora in the digestive tract to act on the feed [49, 52]. However, this hypothesis needs further research as reports seem inconclusive. Further drop in feed intake was recorded with increasing the degree of water restriction in South African indigenous goats [52]; but an improved nutrient utilization was also reported by the same author. Concomitantly, Ahmed Muna and El Shafei Ammar (2001) [49] reported an improvement in digestibility of Lucerne hay under water restriction in desert goats, and similarly, higher organic matter digestibility was observed in water-restricted dairy cows [53]. These adaptations allow the exploitation of grazing areas which are distant from water sources, and prevent erosion especially in regions where water is scarce and grazing pressure is high. In contrast, others [46, 48, 54] found no changes in feed digestibility in water-restricted sheep and goats. It was suggested that the elevated digestibilities usually observed are rather the result of dry matter accumulation rather than "a real increase in fermentation or digestion" [50].

The drop in feed intake puts an additional burden on the water stressed animal. In fact, in order to survive such regimes, adapted ruminants are thought to resort to lowering their metabolic rate in order to reach a new body condition with lower maintenance requirements [55]. Consequently, the effects of this decrease in dietary intake should be considered along with the effects of dehydration. Feed restriction of 50% for only a 3-day period is enough to cause metabolic changes in lactating dairy Sarda ewes [56]. It has also been reported that the depleted body condition during periods of energy deficiency reduces heat tolerance [57], which in turn affects the reproductive potential of sheep [58, 59].

As reported in [9, 47, 60-64] the most obvious physiological consequence of water restriction with the concomitant reduction in feed intake is weight loss. Many trials on dry and lactating Awassi ewes recorded a drop in weight ranging between 0.84% and 26% (Table 1). Besides the effect of the water regimen, other factors contribute to body weight variation such as the physiological status of the animal (lactating or dry), its age, and the prevailing climatic conditions during the experiment (ambient temperature). It is clear in Table 1 that watering every two days did not cause a mentionable weight loss in Awassi ewes even if the temperature reached up to 32°C. The highest weight loss (26.2%) was recorded in young sheep (2-year-old ewes) and in lactating animals. Reported results lead to one conclusion that dry Awassi have a high adaptation to dehydration, and can tolerate 3-day water restriction regime, up to one month with losing only 16.8% of their body weight (Chedid et al., unpublished).

Significant weight loss is documented in other breeds of sheep and goats subjected to feed and water stress [6, 65]. Part of this weight reduction is due to body water losses [9] while the other part is caused by the consequent mobilization of fat (and possibly muscle) used for energy metabolism to compensate the decrease in dietary intake [9, 66]. Furthermore it was observed that water restriction leads to more weight loss as compared to feed restriction alone [49, 62-63] although the difference was not always statistically significant.

The following table (Table 2) presents the effect of water and feed restriction on body weight of adult dry Awassi ewes [62, 63].

Results in both studies are in compliance: water restricted animals lost more weight (approximately two fold) than those drinking every day but receiving less feed; however, large individual variations were recorded within each of the experimental groups (N=8). Therefore, further research is needed for conclusive results on the difference between water and feed restriction impact on weight loss.

Physiological status	Water restriction regime	Drop in weight (%)	Age	Ambient temp. (°C)	Reference
Non-lactating	2-day-restriction	0.84	mature	15-32	[9]
	3-day-restriction	9.98	mature	27-30	[62]
		16.8	mature	23-28	[63]
		16.7	mature	27-31	[47]
		26.2	2 years	30-31	[64]
		10.4	mature	18-21	
	4-day-restriction	3.32	mature	15-32	[9]
	1L on day 4 and 3L on day 8 of 12-day water restriction	22.13	mature	25-35	[61]
	1L on day 4 of 7-day water restriction	16.8	mature	23-33	Chedid et al. (unpublished)
Lactating	3-day-restriction	26.2	mature	27-31	[47]

Table 1. Effect of water restriction on body weight of Awassi sheep.

	Treatment*		Reference
	Water restriction	Feed restriction	
Weight loss (%)	9.98	5.7	[62]
	17.9	8.2	[63]

*Water restricted animals received water every 3 days;

Feed restricted animals had free access to water but received 60% feed of the ad libitum intake.

Table 2. Effect of water restriction versus feed restriction on body weight in Awassi ewes.

3.2. Fat metabolism: Fat cell diameter, cholesterol, glucose, fatty acids, leptin and insulin

The fat deposited during the periods of pasture abundance is mobilized and utilized for maintaining the body and sustaining production during periods of scarcity [67, 68]. The specialized fat depot represented in the fat-tail of many indigenous sheep, serves as a readily available source of energy to circumvent variation in dietary energy intake. This was well described in the Barbarine sheep subjected to long periods of undernutrition [37] as well as in the Awassi which showed a reduction in the fat-tail adipocyte diameter following an intermittent watering regime [64]. Ermias et al. (2002) [68] highlighted the importance of the location of fat depots as adaptive features to periodic fluctuations in nutrition. They noted that the rump and fat-tail depots are the most responsive under such conditions. On the other hand, Atti et al. (2004) [37] noted that subcutaneous fat is the first energy depot to be mobilized when energy intake is deficient. This is true for the fat-tailed sheep such as the Barbarine on which they conducted their study, as well as other ruminants. However, the fat-tail provided an adaptive advantage by being slowly mobilized when undernutrion is extended over a long period thus allowing long-term survival by using up this important energy store.

Fat mobilization under water restriction is further denoted by high levels of cholesterol [69] and Free Fatty Acids (FFA) in the blood [61, 64]. The increase in plasma cholesterol following water scarcity is attributed to the decrease in energy intake, along with fat metabolism [70]. These results were recorded in different ruminant species submitted to water deprivation such as Awassi [9, 47] and Yankasa ewes [8]. Furthermore, the high level of FFA outlined in water restricted Awassi [61, 64] and the Sudanese desert sheep [71] reflects lipid mobilization within the adipocyte, thus permitting lipid stores to be used as fuel when feed intake is limited [72]. Similarly, FFA levels in the blood of lactating goats correlated positively with fat mobilization in times of under nutrition [73].

Reports about the changes in glucose levels in water restricted sheep are contradicting: while no significant change was recorded by some authors [8, 9, 61, 74], a decrease in plasma glucose level was observed in Merino sheep after 24 and 48 hours of fasting [75], and in intermittently watered Awassi, although the change was not significant [47]. Glucose metabolism decreases due to the decline of propionate (the major precursor for gluconeogenesis) production in the rumen caused by low feed intake [76]. Although ruminants derive most of their energy requirements from volatile fatty acids resulting from rumen fermentation of carbohydrates, they still have an absolute need for glucose, necessitating a good homeostatic control of this compound [77].

Fat mobilization seems to be coordinated by underlying changes in the levels of key hormones. In dry Awassi ewes, subjected to an intermittent watering once every four days, FFA were negatively correlated with insulin and leptin as highlighted in [64]. The decrease in plasma insulin is probably caused by the decline in feed intake, insulin secretion being accelerated by feeding [78]. Insulin levels are thought to remain low during fasting periods in order to facilitate lipolysis [79]. Similarly, the decrease in leptin levels in dehydrated and undernourished ruminants is well documented and explained by the decrease of the metabolic status which inhibits the adipose tissue from secreting leptin [80]. Moreover, a strong correlation between leptin concentration and fat-tail adipocyte diameter was noted in [64] highlighting the relation between the secretion of leptin by the adipose tissue and body fatness [81]. Chilliard et al. (2000) [36] proposed a model whereby insulin, cortisol and leptin interact in the process of adaptation to underfeeding and re-feeding in ruminants such as experienced also under intermittent watering. The drop in leptin following undernutrition leads to a chain of events that includes stimulation of re-feeding and decrease of energy expenditure and insulin-sensitivity, that serves to re-establish homeostasis by preventing excessive lipolysis that would yield toxic levels of fatty acids, and preserve body stores to prolong survival [36, 81].

3.3. Hematology: PCV and hemoglobin

Dehydration in warm weather conditions reduces plasma volume as water is taken up by the tissue [82]. Although some authors reported and agreed that increased PCV and Hb concentration are good indicators of dehydration [42, 61, 71, 83, 84] results on these two parameters have been inconclusive. Even though levels of hematocrit were found to increase in Awassi [42, 60] and Merinos [61] subjected to water stress in some experiments, no variation was remarked in Yankasa [7, 8], Awassi [9] and Australian whether sheep [85] under similar watering conditions. Similar contradictory results were reported regarding hemoglobin: whilst an elevation of hemoglobin level was attributed to a decrease in plasma volume due to water loss [47, 61, 85] others did not report any variation [7-9]. These undetermined results may be an indication that adapted sheep can maintain plasma volume [86] and redistribute body water after a long water deprivation period [87].

3.4. Blood chemistry

3.4.1. Total protein, globulin and albumin

According to Caldeira et al. (2007a, b) [88, 89], serum total protein, globulin and particularly albumin are good indicators for predicting the animal's protein status. A drop in serum albumin concentration is observed in ruminants with low dietary protein intake [88, 90], followed by a decrease in globulin concentration when this dietary insufficiency is prolonged [88]. Many authors reported an increase in blood albumin and globulin in sheep under water restriction [9, 61, 74, 91, 92]. The high protein concentration is explained by the reduced plasma volume due to dehydration [93]. However, reduction in total protein and albumin was noticed after 3 days of water deprivation in Awassi [47] and in Barki sheep under water stress [94] suggesting that low feed intake is behind this reduction and that circulating proteins are being used in order to compensate for the dietary shortfall. Accordingly, maximum values of total serum protein were recorded on the 8th day of water restriction in Awassi followed by a decline at the end of the experiment (12 days) [61]. Serum albumin is a major labile protein reservoir, but it is also very important in the maintenance of body osmoregulation. Consequently, some variations in serum albumin levels can occur, but the maintenance of normal levels has to be re-established as soon as amino acids from other sources like the skeletal muscle are available [95].

3.4.2. Creatinine and urea

Urea is mainly synthesized in the liver using NH_4^+, the end product of protein catabolism, and is released to the blood [96]. Urea is excreted by the kidneys to rid the body of the excess N intake that was not used for maintenance or production [90], or it is recycled through saliva or by reabsorption into the rumen to be utilized by rumen microflora [96, 97]. Creatinine is produced in the muscles and excreted by the kidneys in proportion to the muscle mass and the rate of proteolysis [88, 97]. The transfer function of the kidney is altered under water stress [98] with slower glomerular filtration and higher urea re-absorption [6, 8, 99]. Water stress induces a decrease in urine output and the production of dry faeces under the action of vasopressin, and increased water reabsorption from the gastro-intestinal tract [100]. Urine volume dropped by 75% and fecal water output was 37% lower in desert sheep subjected to 5 days of water restriction [101]. Consequently, urea reabsorption by the kidney is also expected to increase as reflected by increased concentration in the blood [102], which was confirmed by several trials on Merinos [65], Yankasa [7, 8] and Awassi sheep [9, 42]. When Yankasa sheep were submitted to two consecutive periods of five-day water stress, an increase in urea and creatinine concentration was observed after the first period but only creatinine levels remained high after the second [8]. Thus the author suggested that urea is being re-circulated from the blood system into the digestive tract. This is consistent with the observations that urea conservation at the level of the kidneys and recycling into the gut is increased when dietary nitrogen intake is low [102].

On the other hand, creatinine levels in lambs were not affected by 48-hour water restriction [103], while others observed an increase of this parameter in water restricted animals [47, 91]. The creatinine concentration is influenced by the level of reliance on proteolysis and endogenous N sources [88, 98] as well as by higher kidney retention due to decreased glomerular filtration rate. In turn, these factors are related to the degree of protein/N intake deficiency that the animal is experiencing as well as the level of dehydration.

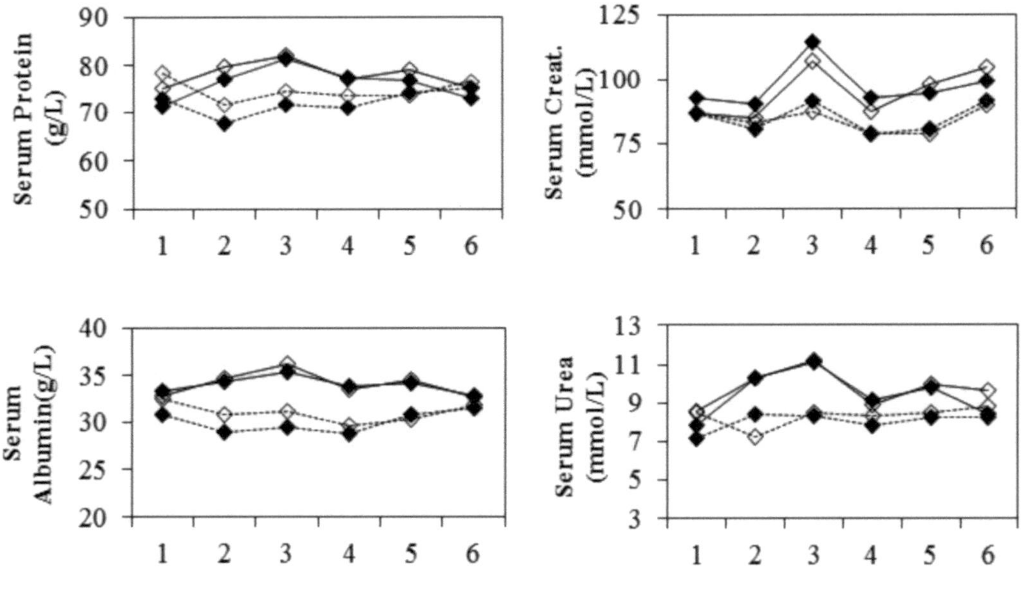

Figure 1. Changes in some serum chemistry indicators of dry (◊) and lactating (◆) Awassi ewes subjected to daily watering (—) or to water restriction (- - - -). (Adapted from [47])

Figure 1, adapted from [47], illustrates the changes over repeated cycles of three days water restriction on major blood chemistry parameters of dry Awassi ewes. This figure shows that the response to the watering regime tends to decrease over time after it reaches a certain peak. This underlines the adaptation mechanisms that the animals activate in order to achieve homeostasis under the imposed treatment. The observed return to normal values in protein and albumin concentrations could indicate that these compounds are being used up from the blood in replacement of the deficient dietary intake, or this could also denote the mobilization of water from the extracellular fluid and the rumen into the blood stream [47]. Similarly, the decline in urea at the end of the treatment could be related to the above mentioned urea recycling mechanism, while creatinine concentrations remained relatively high as previously observed in [8]. The figure also illustrates that lactating and dry animals had similar responses to water restriction with no effect of lactation (2-3 months in lactation) on these blood parameters.

3.4.3. Electrolytes and osmolality

Analyzing osmolality is a good approach for monitoring the hydration status. It has been proved that osmolality and electrolytes levels are largely affected by water deprivation: the reduced plasma volume causes hyperosmolality inducing consequently an increase in electrolytes concentration [104] mainly sodium Na^+ and chloride Cl^- [60, 105, 106]. These results were reported in different breeds of sheep and goats [10, 63, 71, 91, 103, 107].

Increased renal retention of Na^+ is a physiological response to water restriction in different small ruminant breeds, which allows the maintenance of sodium balance in the body. Ashour and Benlemlih (2001) [87] attributed the increased renal retention to the influence of aldosterone, whereas McKinley et al. (2000) [107] added the effect of vasopressin secretion. Dehydration causes an increase in plasma vasopressin levels in both lactating goats [108-110] and non-lactating ones [108, 111]. According to Yesberg et al. (1970) [112] urinary vasopressin excretion rate is directly related to urinary osmolality and inversely related to urine flow rate. This explains why dehydrated goats and sheep decrease their urine volume while the osmolality and vasopressin levels augment. Olsson (2005) [113] noted that thermoregulation and fluid balance, as regulated by thirst control, vasopressin secretion, sodium balance and other osmotic and cardiovascular signals, seem to be centrally regulated at the hypothalamic level, particularly the preoptic and anterior hypothalamic neurons. Silanikove (1994) [15] presents a detailed account of the dynamics of major electrolytes under dehydration in connection with water conservation and homeostatic mechanisms in ruminants. In addition to the increased renal retention of water and Na^+, saliva secretion is decreased in dehydrated animals but its osmolality is increased. In parallel, the ruminants resort to utilize the large volume of water present in the gut through active transport of Na^+ across the rumen wall. This transport necessitates the presence of a minimal amount of volatile fatty acids in the rumen, hence the importance of sustaining some feed intake during dehydration. The hyperosmotic fluid absorbed from the rumen needs to be desalted the salivary flow to the rumen has to be maintained in order to preserve homeostasis.

Rehydration is an equally challenging situation for ruminants that can lead to hemolysis in non-adapted or severely dehydrated animals. Many studies reported the slow return to normal levels of blood volume, osmolality and other blood components after the rehydration of ruminants, although the animals drank large amounts of water at once [42, 92, 114]. The mechanisms allowing the slow release of the ingested water from the rumen into the blood stream are not fully understood. The production of hypotonic saliva is dramatically increased following rehydration [15] thus allowing the recycling of absorbed water and Na^+ from the blood back into the rumen to prevent a sudden drop in blood osmolality. In parallel, the kidney sustains its water/Na^+ conservation activity to prevent the loss of the ingested water which is vitally needed in anticipation of a future dehydration cycle. Rehydration also activates appetite and thermoregulatory mechanisms that allow the final restoration of homeostasis and normal functioning in up to 24 hours after rehydration or more.

Studies on different sheep breeds [115, 116] showed a negative correlation between Na^+ and K^+ in plasma. Concurrently, blood K^+ was reported to decrease in water-deprived sheep [9, 65] probably due to the intra-erythrocytic diffusion of K^+ or loss of these ions in urine in exchange of Na^+ re-absorption [8]. However, others observed an elevation in plasma K^+ under water restriction [7, 117], while [8, 47, 61] did not report a variation in K^+ levels in Yankasa and Awassi sheep, respectively. These inconclusive results about K^+ alteration under water restriction do not make of potassium a reliable indicator of the hydration status; on the contrary they warrant further studies about the role played by this blood parameter during dehydration.

Chloride Cl^- is the major anion in extra cellular fluids (ECF). It functions primarily in transport processes integral to cation and water balance and as a conjugate anion in acid-base metabolism. Several findings reported that dehydration leads to an increase in plasma chloride levels in parallel to sodium levels [8, 9, 47, 74] as Cl^- is passively distributed in relation to the electrical gradients established by active Na+ transport [118]. This increase may be attributed to many phenomena such as the hemoconcentration resulting from a lower blood water level [74], and the increase in aldosterone and vasopressin concentrations [87] leading to increased renal retention.

Calcium plays an important role in regulating ion gating and as a co-factor for intermediary metabolism reactions. However, studies did not report variation in Ca^{++} under water deprivation in the Awassi [9, 47] nor in Comisana sheep [74].

Finally, blood pH, a critical parameter for normal enzymatic and metabolic functions, seems to be well maintained in intermittently watered Awassi [9, 47, 61]. Increase in pH was only recorded in highly restricted Awassi following a once in a five days intermittent watering regime [9], this could be related to a combination of the high dehydration state and environmental heat which leads to hyperventilation and consequently to respiratory alkalosis as observed in other heat stressed animals due to the increased elimination of CO_2 [119].

Figure 2 is an illustration of the effect of a 12 days water restriction episode on Awassi dry ewes. The animals were offered 1 liter of water on day 4 and 3 liters on day 8 only. In this trial, the treatment had no significant effect on blood pH and K^+ concentration, while Na^+ and Cl^- were significantly increased under the restriction regime. The trial also included a group of water restricted animals that received 2.5g/d of vitamin C, the effect of which is discussed below (vitamin C section). It is worth noting here that the levels of Na^+ and Cl^- seem to have reached a peak after which they started to show a slight decline. This observation reinforces the previously proposed idea of adaptation to the restriction regime to reach a new homeostasis by activating water mobilization and conservation mechanisms to restore the blood volume and composition.

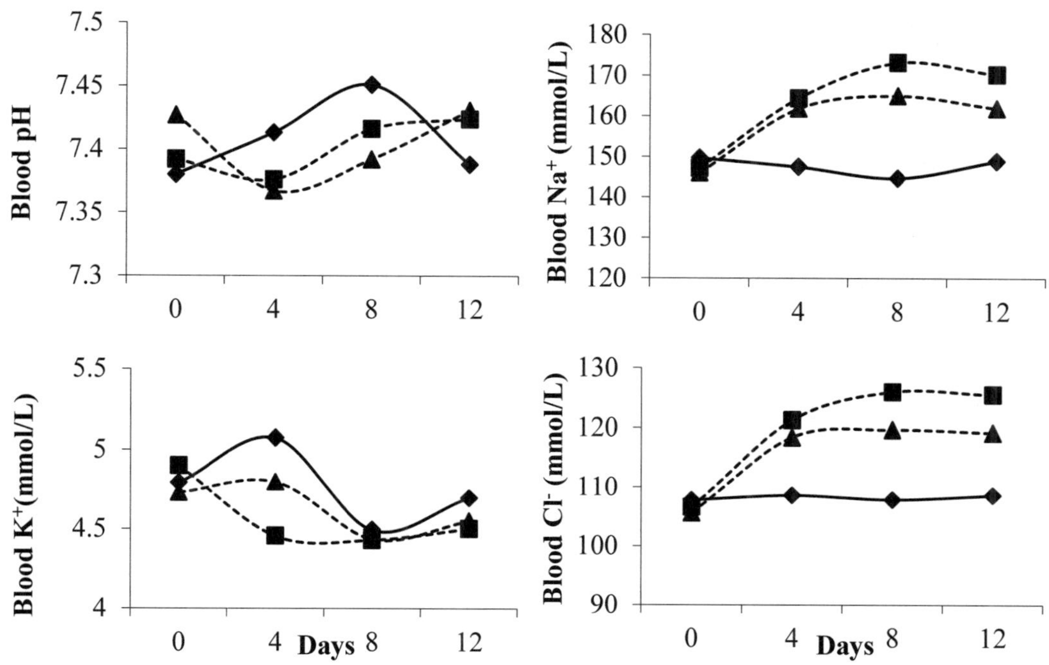

Figure 2. Effect of water restriction (- - -) on dry Awassi ewes with (▲) and without (■) vitamin C supplementation (Ghanem et al., 2008, unpublished data)

3.5. Cortisol and other hormones

Cortisol is a hormone secreted in order to deal with stress. It is released due to the activation of the hypothalamo-pituitary-adrenal axis by stress. Although it plays a major role in maintaining the balance of water and electrolytes [120, 121], its mechanism is not very clear yet [122]. Dehydration had no effect on serum cortisol levels of Awassi [9, 61] and Clun forest sheep [123] which is consistent with the results obtained under laboratory conditions in sheep deprived of feed and water for 48h [124]. On the other hand, Kataria and Kataria (2004) [98] suggested that the increase in cortisol levels in dehydrated Marwari sheep (for 6 days) is a sign that the animals are under stress; they also reported that cortisol levels did not return to normal even after 72 hours of rehydration. Working on the same subject, Li et al. (2000) [85] reported plasma cortisol variation with fasting while water restriction had no additional effect on this parameter. A decline in serum cortisol was recorded in intermittently watered Awassi along the experimental period [47]. Concurring with [125], it is suggested that cortisol could be a good parameter in assessing acute stress response in small ruminants but not chronic stress such as dehydration.

Thyroid hormones (Triiodothyronine T3 and Thyroxine T4) play a major role in many physiological events such as thermoregulation and metabolic homeostasis of energy and proteins [126-128]. T3 and T4 concentration is affected by many factors like reproductive status, climatic conditions [127, 129] nutrition, age and gender [130, 131]. Water restriction [64] and nutrient limitation [132] were found to lower the levels of T3 and T4 in dry Awassi ewes and pregnant Whiteface Western ewes, respectively. Similarly, Caldeira et al. (2007a) [88] noticed a decrease in these hormones with decreasing body score of ewes. They concluded that T3 is a good indicator of the metabolic state of the animal. On the other hand, variable T3 and T4 responses to seasonal variations and/or dehydration were reported in literature from various ruminants [133]. The concentrations of these two hormones were found to be strongly correlated (R^2=0.568; P=0.000) [64]. The authors suggested that this decline in T3 under water restriction reflected the declining metabolic state due to dehydration and decreased feed intake while the declining T4 concentration was probably a response to the thermal stress experienced by the animals in that experiment. Similarly, T4 concentrations were reported to vary with the season while T3 is affected by both the season and the physiological status of the sheep [134]. The reduction of thyroid hormone activity under dehydration is associated with the animal's attempt to minimize water losses by reducing general metabolism [135].

4. Changes in relation with physiological status

As seen above, small ruminants in arid and semi-arid regions face many constraints related to fluctuating temperatures as well as shortages in feed and water sources. Pregnancy and lactation increase the needs for adaptive mechanisms due to the greater need for food, water and electrolytes in order to meet the requirements of the fetus and the mammary glands [113]. Water requirements could increase up to 50% by late pregnancy (around 160 ml/kg $BW^{0.75}$) while the requirement for milk production is stated as 165 ml/kg $BW^{0.75}$ for a milk production level of 148g milk/$kg^{0.75}$ [5]. Pregnant and lactating animals have 40-50% higher water turnover rates than dry animals [17]. However, Degen (1977) [35] recorded a small difference in water turnover between control and pregnant Awassi and Merinos sheep.

4.1. Pregnancy

The reported physiological changes vary according to the degree of water restriction and stage of pregnancy. Chokla pregnant ewes submitted to intermittent watering (every 72 and 96 hours) under semi-arid conditions showed significant hemocentration and reduction in extracellular fluid space as compared to pregnant Chokla receiving water daily [136]. Similarly, Olsson et al. (1982) [137] reported increases in plasma osmolality and Na^+ concentration in pregnant goats dehydrated for 30h accompanied by a decrease in glomerular filtration rate, while plasma protein and hematocrit did not change with dehydration. Interestingly, it was observed that pregnant goats [137] and sheep [138] have a lower capacity to concentrate urine in response to dehydration. Authors in both studies suggested that the apparent reason for this observation is a decreased sensitivity to Arginin-Vasopressin (AVP) which in turn, could be partially due to the effects of high prostaglandin concentrations which increase during late gestation. Research on the Matebele goat showed that a low nutrition level during late pregnancy had little effect on kid birth weight [139]. Similarly, a twice weekly watering regime imposed for a prolonged period had no effect on birth weight of desert adapted Magra and Marwari sheep [140]. On the other hand, pregnant Chokla ewes watered every 4 days gave birth to lambs of lower weight compared to ewes that were watered daily or every three day; however, at 12 weeks of age, lambs' weight were similar between the differently watered groups [136]. Working on goats, Mellado et al. (2006) [141] highlighted the importance of goat birth weight and weight gain at 25 days of age on their future reproductive performance under intensive conditions in hot arid environments. On the other hand, in a recent study prenatal feed restriction in the last trimester resulted in lower male offspring weight in goats, but had no effects on later behavior and growth [142]. Further studies are needed to assess the long term consequences of dehydration and/or intermittent watering during gestation on the growth and later performance of the offspring.

4.2. Lactation

Significant weight loss was recorded in lactating water-stressed Awassi [47] and Comisana ewes [74] as compared to control animals. Weight loss during lactation is due to body water loss caused by less water and feed intake combined with energy deficit that drives lactating animals to strongly depend on their body reserves [143].

Physiologically, lactating animals show lower hemoglobin concentration than dry ones [47]. According to [120], the decrease in hemoglobin concentration in lactating ewes is explained by the high water content and plasma volumes due to increased water mobilization to the mammary glands. It was found in [47] that lactation did not affect levels of serum glucose or those of cholesterol in Awassi ewes in their second to third month of lactation. Concurrently, no significant changes were recorded in glucose level caused by lactation beyond the first month [144]. Moreover, lactation did not affect blood total protein, albumin and globulin concentration of Awassi ewes in their mid-lactation [47]; no change in albumin was also reported in [145], however an increase in the gammaglobulin fraction was noticed. El-Sherif and Assad (2001) [94] observed a return to normal total protein levels on the fourth week of lactation most probably caused by a fall in globulin concentrations. The same authors reported that lactation significantly increased plasma albumin, albumin to globulin ratio and blood creatinine in Barki ewes under semi-arid conditions. No alteration was caused by lactation on serum urea and creatinine levels in Awassi ewes [47] and Corriedal [138]. Working with lactating Comisana subjected to 60% water deprivation, Casamassima et al. (2008) [74] reported significant elevations in serum concentration of triglycerides, albumin, total proteins and cholesterol. Intermittently watered Ethiopian Somali goats, exhibited similar physiological responses to those described in other breeds, namely increased osmolality, AVP secretion and blood protein concentration [110]. The authors also noted a significant capacity to fluctuate rectal temperature in response to heat stress reaching 5°C daily change in some animals; they also observed the activation of a water saving mechanism following the first cycle of water restriction, resulting in lower physiological changes in subsequent cycles. A similar trend was also noted in lactating Awassi ewes subjected to intermittent watering [47].

Concerning the effect of lactation on blood pH, Hamadeh et al. (2006) [47] reported significantly higher blood pH in water restricted lactating Awassi ewes as compared to dry ones. This increase in pH is correlated to the decrease in plasma Ca^{++} and K^{+} levels due to their need for milk production [47] and to the increase in plasma Na^{+} and Cl^{-} since Na^{+} is used for nutrient transport [146]. As for cortisol, lactating Awassi ewes tended to have high levels as compared to dry counterparts. However, more work is needed in order to elucidate the adaptive mechanisms of pregnant and lactating ewes to water restriction.

4.3. Milk production and composition

Water needs are reported to be the highest during lactation as compared to other physiological statuses [109]. Dehydration leads to a reduction in the blood flow to the mammary gland however, enough supply for milk production may still be achieved as supported by the sustained milk production in dehydrated goats [113] and sheep [74]. During dehydration, milk volume is generally decreased [109, 110, 147], although the Bedouin goats could maintain their milk production when watered every second day [108]. Dahlborn et al. (1997) [148] suggested that the drop in milk volume observed in some dehydrated animals was mainly the result of lack of water as such and not a reflection of the consequent decrease in feed intake; furthermore the authors suggested that this drop could be related to the alteration in casein production observed under dehydration. In fact, Silanikove (2000) [6] showed that stress leads to a chain of events including increased cortisol secretion leading the activation of the plasmin system resulting in the release of a proteose peptone with channel blocking activity (PPCB) from β-casein and interfering with lactose secretion into the lumen of the mammary gland and consequently causing a drop in production. Previous studies have indicated an increase in milk osmolality, lactose and density following water deprivation for 48 hours [109, 149, 150]. Milk osmolality is strictly controlled to keep it isotonic with plasma; the increase in lactose under water restriction, being the major osmotic component of milk, is probably a response to the increase in serum osmolality under water restriction [149]. Nonetheless, milk can be less concentrated in fat and non-fat solids and more rich in water. Alamer (2009) [150] noted a decrease in fat content of milk of 25% water restricted goats but not in those that were 50% restricted. The increase in water content [151, 152] is thought to be a form of adaptation allowing the offspring to receive the adequate quantity of water when water is not available.

5. Stress alleviation drugs

Domestic animals are routinely faced with different stressors. Most stressful conditions, including diseases [153] and farming practices such as milking [143], isolation [154], introduction to a new flock [155], road transportation [156, 157] last for only a short period ranging between hours to days. On the other hand, harsh environmental factors are probably the stressors with the longest lasting effect since they may prevail for months. For example, heat stress and elevated ambient temperature are considered major risks affecting sheep performance [85; 158]. The negative impact of heat is translated into increased body temperature, higher respiration and heart rates followed by a drop in feed intake, redistribution in blood flow and alteration in endocrine function [158]. Low temperatures or cold leads to an equally stressful situation which affects sheep performance by increasing the metabolic rate [85,160]. Furthermore, different environmental constraints often come together such as the situation in arid and semi-arid areas during the dry season when heat stress is combined with water scarcity and low pasture quality.

Animal producers and researchers have looked for ways to alleviate the negative effects of common stressors. Stress alleviation strategies are numerous, and their availability to producers depend on the access to water and energy, the price they are able to pay and the adopted farming system [153]. These strategies vary from simple on-farm practices such as modifying the feeding pattern, feed composition [160], water management, cooling systems and environmental modifications like shading [153, 161], protection from solar radiation [153], the use of micro-sprinklers, spray jets and ventilation [162] to more scientific procedures like genetic selection [161] and others.

In this quest for stress alleviation in domestic animals, researchers have tested special drugs and/or nutritional supplements. Trials showed that a pre-transportation administration of ascorbic acid to goats facilitates the transition from depression to excitation; it exhibited potential depression amelioration after road trips [163] and significantly decreased weight loss caused by transportation under unfavorable thermal conditions [164]. Ali et al. (2005) [156] reported that a single dose of the anti-stressor xylazine administered to sheep and goats before road transportation considerably ameliorated the effects induced by the stressful stimulus; whereas a pretreatment with sodium betaine (a test compound) had no significant effects. Electrolyte therapy was also found effective in reducing stress of transportation in market cattle allowing a better meat quality and a reduction in live weight loss [157].

In this review we will focus on two compounds that appear to have a good potential for stress remediation in domestic animals: Vitamin C which is tolerated at high doses without apparent side effects [165] and aspirin which showed some potential advantages warranting further investigation [166-168].

5.1. Vitamin C

Although ruminants biosynthesize ascorbic acid under normal conditions and do not need any additional supplementation [45], and even though vitamin C administration is not a common practice in adult livestock nutrition [169] scientists however, decided to study the effect of Vitamin C administration to sheep under water-stress conditions [60, 61, 64, 92] and goats facing the stress of transportation [163, 164]. These trials were encouraged by promising results obtained on weaned pigs [170], Japanese quails [171], rabbits [172] and broilers [173] under stress.

Ascorbic acid is known for its function as an antioxidant mainly due to its redox properties; it acts as a free radical scavenger in numerous cellular oxidation processes [174] and has been demonstrated to be helpful for young ruminants in acclimatizing to cold stress [175]. Ascorbic acid plays an important role in modulating the immune response by enhancing neutrophil function and minimizing free radical damage [176] and by improving antibody response to antigen [177]. Concurrently, Minka and Ayo (2007) [164] reported that administering Vitamin C to Red Sokoto goats before transportation reduced the post-journey effect to a minimum or eliminated it completely; however the impaired animals' homeostasis was rapidly recovered after the trip.

A daily dose of Vitamin C was reported to decrease weight loss in adult female Awassi subjected to a water-restricted regime [61, 63] while this effect was not significant in other trials [63, 64, 92]. The alleviation of body weight loss can be explained by improved feed intake [92] and better feed conversion, also observed in other animals in stressful conditions and supplemented with vitamin C [164, 171, 178]. The ameliorated effect of ascorbic acid on weight loss during short-term transportation of goats in hot weather [164, 179] proves the advantage of Vitamin C supplementation in order to maintain an adequate live body weight for slaughter.

Vitamin C supplementation to sheep alleviated the effect of dehydration on PCV [61] but not on hemoglobin [61, 63]; while in goats submitted to transportation under unfavorable climate conditions, Vitamin C significantly decreased levels of both PCV and Hb [164]. It was also found that supplementing Vitamin C to deficient pigs increased hemoglobin levels probably due to increased iron absorption [180].

Lower serum protein concentrations were reported in Vitamin C administered water-restricted Awassi as compared to non-supplemented counterparts [61], while others found no significant differences in total protein and globulin levels due to Vit C [92]. The effect of vitamin C on albumin levels is inconclusive as well: although some [61] noted lower concentrations in treated ewes, others [47] reported higher values of albumin in treated animals. The daily Vitamin C dose also plays a considerable role on blood parameters: a daily dose of 5g significantly increased serum creatinine and urea concentrations as compared to 3g and to control [92]. This might be considered as an enhancement to the adaptive mechanisms of Awassi sheep to water restriction. However, in other experiments no effect of Vitamin C was observed on creatinine levels [63] warranting further investigations on the role of Vitamin C in urea and creatinine dynamics during dehydration.

In an experiment conducted in order to study the effect of Vitamin C on fat mobilization under water stress, Jaber et al. (2011) [64] reported no significant effect of this drug administration on adipocyte diameter, fat mobility and weight loss in water deprived sheep. Authors speculated that a daily dose of Vitamin C (3g or more) may be more helpful in increasing fat mobilization under water stress than single high dosages; they agreed that more work would be essential to confirm the observed trends.

A tendency for higher cholesterol levels was observed under Vitamin C administration [60, 61, 63]. Vitamin C interferes in norepinephrine formation, an important hormone that increases fat mobilization [172], it is also essential in carnitine formation, which upon reacting with acetyl CoA forms acetylcarnitine that transports fatty acids into the mitochondria to be oxidized [60].

Reports on the effect of Vitamin C on osmolality are scarce. However, Karnib (2009) [63] and Hanna (2006) [106] observed increased osmolality in water-stressed vitamin C supplemented Awassi ewes, but the mechanism of such phenomenon is still not well understood. Results obtained on the effect of Vitamin C on blood electrolytes are not very clear and need more elaborated work. Ghanem et al. (2008) [60] reported that Vitamin C administration alleviated the effect of water restriction as reflected in lower Na^+ and Cl^- (Figure 2). The authors attributed this observation to the role of vitamin C in norepinephrine formation, which affects the kidney function and therefore water and electrolytes dynamics. On the other hands, [72] reported an opposite result in one experiment while in a second experiment Na^+ and Cl^- were the same between supplemented and un-supplemented water restricted ewes.

Benefits of Vitamin C in decreasing stress hormones was reported by several authors: some [181] observed that ascorbic acid intake resulted in a drop in adrenal and plasma corticosterone levels, others showed that vitamin C eliminated the secretion of cortisol in animals subjected to stress [182]. Still others did not observe any effect of Vitamin C administration on cortisol concentrations in stressed animals [61, 122, 124]. Serum cortisol is a better marker for acute stress than for chronic stress [125]. While Vitamin C and cortisol interact, the anti-cortisol role of Vitamin C is still unclear and needs more research.

Concluding with Vitamin C, it has been shown that the most important parameter in highlighting the role of Vitamin C in counteracting the effect of water deprivation on sheep is the observed decrease in weight loss. Consequently, further work is warranted in order to elucidate the mechanisms of action of Vitamin C, and to determine the best dose recommendations that would increase the adaptive capacities of shepherds and flock keepers to the changing weather and increasing global warming.

5.2. Aspirin

Acetyl-salicylic acid ASA or aspirin has been used for ages as an antipyretic and analgesic agent [183]. New studies have emphasized the role of aspirin in the treatment of some types of cancer [168] and cardio-vascular diseases [184] although data on the effect of aspirin in animal production is contradictory. On one hand, supplementation of 20ppm of ASA to layer chicken under hot climates improved the number of eggs as well as their weight; it increased feed intake, improved fertility and hatchability [185]. On the other hand, chronic feeding of ASA had detrimental results to layer breeders with concerns for early hen livability and egg quality [186]. In mammals, aspirin reduced scouring and improved growth rate when supplemented to weanling pigs at a level of 125 or 250 ppm [166]. It also reduced plasma cholesterol levels in rats [167], and protected rats from colon cancer [168]. Recently, the role of aspirin in the protection from oxidative stress has been highlighted.

The role of aspirin in alleviating stress has been investigated in adult Awassi ewes subjected to water and feed restriction has been studied [62]. Treated animals with a daily dose of

100mg of ASA lost more weight than the untreated animals. However, the difference was not statistically considerable. Similar results had been reported in broilers [187] and weanling pigs [166].

Aspirin did not have any effect on rectal temperature; perhaps these results highlight the capacity of Awassi to remain thermostable even under dehydration [10, 35]. No changes were detected in PCV and hemoglobin levels [62] confirming that salicyclates do not usually alter these two parameters [188]. Moreover, results of the experiment revealed that aspirin has no effect on plasma concentrations of proteins, globulin and albumin. Similarly, no significant differences in the levels of urea, creatinine and osmolality were observed between treated and untreated animals. Furthermore, no significant effect of aspirin was observed on any fat mobilization indicator i.e. cholesterol, insulin, free fatty acids and leptin.

Other studies investigating the role of aspirin in water and feed stressed ruminants are not found. The available literature covers stress resulting from transportation and exposition to new environment [189] and physical pain caused by tail docking [190].

Further studies using different doses of aspirin could be done. Additional experiments might also help clarify the antioxidant property of aspirin and its effect on the ruminant immune system which might be compromised under stress.

6. Conclusion

This review helped in highlighting the adaptability of indigenous small ruminant breeds to water stress and the changes it induces under different physiological statuses. Most small ruminants respond to water stress by decreasing their feed intake, resulting in weight reduction due to water and body mass loss. The rumen plays an important role as water reservoir both in times of dehydration, to maintain blood volume, and upon rehydration to prevent hemolysis. Similarly, modulating saliva production and osmolality is an important mechanism for facing dehydration and rehydration cycles.

A ten-year research track on the Awassi demonstrated the key mechanisms that this breed activates in facing water stress. Strong water conservation is achieved at the level of the kidney, as reflected by a drop in urine output and increased blood Na^+, albumin and urea along with hyperosmolality. Furthermore, the Awassi seems to adapt to an intermittent watering regime after a couple of cycles by re-adjusting its blood volume and constituents to a new status tending toward control levels. Finally, the Awassi mobilizes its fat stores, including the fat-tail, to overcome the shortfall in dietary energy intake. This breed could be maintained, during the hot months and in times of severe water shortage, on short intermittent watering regimes. However, severe dehydration will ultimately result in detrimental effects on milk production, reproductive success, lambs' weight gain and disease resistance, particularly during gestation and peak lactation.

Stress alleviating supplements such as vitamin C show some promise in decreasing the effects of dehydration. Many tracks still need to be explored in future research such as feeding and nutritional manipulations to alleviate water stress, with special attention to the long-term effects of such approaches on overall productivity and welfare. The identification of breeds that show high adaptability to arid and semi-arid regions with an acceptable level of productivity is also important.

Author details

Lina Jaber, Mabelle Chedid and Shadi Hamadeh*

*Address all correspondence to: shamadeh@aub.edu.lb

Department of Animal and Veterinary Sciences, Faculty of Agricultural and Food Sciences, American University of Beirut, Riad el Solh, Beirut, Lebanon

References

[1] Migongo-Bake W. Rumen dry-matter digestive efficiency of camels, cattle sheep and goats in a semi-arid environment in eastern Africa. Rome, Italy: FAO, The complementarity of feed resources for animal production in Africa; 1992.

[2] Iñiguez, L. Small ruminant breeds in West Asia and North Africa. ICARDA Caravan 2005; Issue 22. http://www.icarda.org/Publications/Caravan/Caravan22/Focus_3.htm (accessed 15 August 2012).

[3] Silanikove N. Effects of water scarcity and hot environment on appetite and digestion in ruminants: a review. Livestock Production Science 1992;30: 175-194.

[4] Barbour E, Rawda N, Banat G, Jaber L, Sleiman FT, Hamadeh S. Comparison of immunosuppression in dry and lactating Awassi ewes due to water deprivation stress. Veterinary Research Communications 2005;29(1) 47-60.

[5] Giger-Reverdin S, Gihad EA. Water metabolism and intake in goats. In: Morand-Fehr P. (ed.) EAAP 1991: Goat Nutrition: proceedings of EAAP, 24–26 April 1991, Pudoc Wageningen, Wageningen.

[6] Silanikove N. The physiological basis of adaptation in goats to harsh environments. Small Ruminant Research 2000;35: 181– 193.

[7] Aganga AA, Umunna NN, Oyedipe EO, Okoh PN. Influence of water restriction on some serum components in Yankasa ewes. Small Ruminant Research 1989;2: 19–26.

[8] Igbokwe IO. Haemoconcentration in Yankasa sheep exposed to prolonged water deprivation. Small Rumin. Res. 1993;12: 99–105.

[9] Jaber LS, Habre A, Rawda N, AbiSaid, M, Barbour EK, Hamadeh SK. The effect of water restriction on certain physiological parameters in Awassi sheep. Small Ruminant Research 2004;54: 115– 120.

[10] MacFarlane W.V. Terrestrial animals in dry heat: ungulates. In: Dill D.B., Adolph E.F.A., Wilberg C.C. (eds) Handbook of Physiology, Section 4: Adaptation to the environment. American Physiology Society; 1964. p509-539.

[11] Farid MF, Shawket SM, Abdel-Rahman MHA. Observations on the nutrition of camels and sheep under stress. In: Proceedings of workshop on camels, International Foundation of Science, Sweden, Kartoum, Sudan, 1979.

[12] Turner JC. Osmotic fragility of desert bighorn sheep red blood cells. Comparative Biochemistry and Physiology 1979;A(64A) 167-175.

[13] Ahmed MM, El Kheir IM. Thermoregulation and water balance as affected by water and food restrictions in Sudanese desert goats fed good-quality and poor-quality diets. Tropical Animal Health and Production 2004;36(2) 191-204.

[14] Khan MS, Ghosh PK, Sasidharan TO. Effect of acute water restriction on plasma proteins and on blood and urinary electrolytes in barmer goats of the rajasthan desert. Journal of Agricultural Science Cambridge 1978;2(2) 395-398.

[15] Silanikove N. The struggle to maintain hydration and osmoregulation in animals experiencing severe dehydration and rapid rehydration: the story of ruminants. Experimental Physiology 1994;79: 281-300.

[16] Morand-fehr P. Recent developments in goat nutrition and application: A review. Small Ruminant Research 2005;60(1–2) 25–43.

[17] Cain JW III, Krausman P, Rosenstock A, Turner J. Literature review and annotated bibliography: Water requirements of desert ungulates; 2005. Southwest Biological Science Center.

[18] Miller GD, Cochran MH, Smith EL. Nighttime activity of desert bighorn sheep. Desert Bighorn Council Transactions 1984;28: 23-25.

[19] Langhans W, Scharre E, Meyer AH. Changes in feeding behavior and plasma vasopressin concentration during water deprivation in goats. Journal Veterinary Medicine 1991;A(38) 11-20.

[20] Lechner-Doll M, von Engelhardt W, Abbas HM, Mousa L, Luciano L, Reale E. Particularities in forestomach anatomy, physiology and biochemistry of camelids compared to ruminants. In: Tisserand JL (ed) Elevage et alimentation du dromadaire–Camel production and nutrition. Options Méditerranéennes CIHEAM, Paris 1995;B(13): 19–32.

[21] Skinner JD, van Jaarsveld AS. Adaptive significance of restricted breeding in southern African ruminants. South African Journal of Science 1987;83: 657-663.

[22] Blanc F, Bocquier F, Debus N, Agabriel J, D'hour P, Chilliard Y. La pérennité et la durabilité des élevages de ruminants dépendent des capacités adaptives des femelles. INRA Production Animale 2004;17: 287-302.

[23] Al-Azraqi AA. Effect of fasting on luteal function, leptin and steroids concentration during oestrous cycle of the goat in natural photo-status. Animal Reproduction Science 2007;98: 343-349.

[24] Kosior-Korzecka U, Bobowiec R, Lipecka C. Fasting-induced changes in ovulation rate, plasma leptin, gonadotropins, GH, IGF-I and insulin concentrations during oestrus in ewes. Journal of Veterinary Medicine Series A – Physiology, Pathology and Clinical Medicine 2006;53: 5–11.

[25] Marai IFM, El- Darawany AA, Fadiel A, Abdel-Hafez MAM. Reproductive performance traits as affected by heat stress and its alleviation in sheep. Tropical and Subtropical Agroecosystems 2008; 8(3): 209-234.

[26] Hamadeh SK, Shomo F, Nordblom T, Goodchild A, Gintzburger G. Small ruminant production in Lebanon's Beka'a Valley. Small Ruminant Research 1996;21(3) 173-180.

[27] Amoah EA , Gelaye S, Guthrie P, Rexroad Jr. CE. Breeding season and aspects of reproduction of female goats. Journal of Animal Science 1996;74: 723-728.

[28] Hofmann RR. Evolutionary steps of ecophysiological adaptation and diversification of ruminants: a comparative view of their digestive system. Oecologia 1989;78: 443-457.

[29] Gordon IJ. Browsing and grazing ruminants: are they different beasts? Forest Ecology and Management 2003;181: 13–21

[30] Jonsson H. *Foraging behaviour of cattle, sheep and goats on semi-arid pastures in Kenya. Veterinary Medicine thesis. Swedish University of Agricultural Sciences SLU; 2011. p.1-19.*

[31] Shipley LA. Grazers and browsers: how digestive morphology affects diet selection. In: Launchbaugh K.L., Sanders K.D., Mosley J.C. (ed). Grazing behavior of livestock and wildlife. Idaho Forest, Wildlife and Range Experiment Station Bulletin No. 70; 1999. p.20-27.

[32] Eyal E. Shorn and unshorn Awassi sheep. Body temperature. Journal of Agriculture Sciences 1963; 60: 159-176.

[33] Kay RNB. Responses of African livestock and wild herbivores to Drought. Journal of Arid Environments 1997;37: 683–694.

[34] Degen AA, Shkolnik A. Thermoregulation in fat-tailed Awassi, a desert sheep, and in German Mutton Merino, a mesic sheep. Physiological Zoology 1978;51: 333–339.

[35] Degen AA. Fat-tailed Awassi and German Mutton Merino sheep under semi-arid conditions III: body temperature and panting rate. Journal of Agricultural Science 1977;89: 399-405.

[36] Chilliard Y, Ferlay A, Faulconnier Y, Bonnet M, Rouel J, Bocquier. Adipose tissue metabolism and its role in adaptations to undernutrition in ruminants. Proceedings of the Nutrition Society 2000;59: 127-134.

[37] Atti N, Bocquier F, Khaldi G. Performance of the fat-tailed Barbarine sheep in its environment: adaptive capacity to alternation of underfeeding and refeeding periods. A review. Animal Research. 2004;53: 165-176.

[38] McNab BK. The physiological ecology of vertebrates. Comstock Publishing Associates, Cornell University Press, Ithaca, New York, USA; 2002.

[39] Zervanos SM. Renal structural adaptations among three species of peccary. The Southwestern Naturalist 2002;47(4) 527-531.

[40] Horst R, Langworthy M. Observations on the kidney of the desert bighorn sheep. Anatomical Record 1971;2: 343.

[41] Turner JC. Water, energy and electrolyte balance in the desert bighorn sheep, Ovis Canadensis. Ph. D. Thesis. University of California, Riverside, California, USA; 1973. p.1-276.

[42] Laden S, Nehmadi L, Yagil R. Dehydration tolerance in Awassi fat-tailed sheep. Canadian Journal of Zoology 1987;65: 363–367.

[43] Forbes JM. The water intake in ewes. British Journal of Nutrition 1997;22: 33-34.

[44] More T, Howard B, Siebert BD. Effect of level of water intake on water, energy and nitrogen balance and thyroxine secretion in sheep and goats. Australian Journal of Agricultural Research 1983;34: 441-446.

[45] National Research Council of the National Academies. Nutrient requirements of small ruminants : sheep, goats, cervids, and New World camelids / Committee on Nutrient Requirements of Small Ruminants, Board on Agriculture and Natural Resources, Division on Earth and Life Studies, National Research Council of the National Academies; 2007.

[46] Hadjigeorgiou I, Dardamani K, Goulas C, Zervas G. The effect of water availability on feed intake and digestion in sheep. Small Ruminant Research 2000;37(1/2) 147-150.

[47] Hamadeh SK, Rawda N, Jaber LS, Habre A, Abi Said M, Barbour EK. Physiological responses to water restriction in dry and lactating Awassi ewes. Livestock Science 2006;101(1-3) 101-109.

[48] Maloiy GMO, Kanui TI, Towett PK, Wambugu SN, Miaron JO, Wanyoike MM. Effects of dehydration and heat stress on food intake and dry matter digestibility in East African ruminants. Comparative Biochemistry and Physiology 2008;A(151): 185–190.

[49] Ahmed Muna MM, El Shafei Ammar I. Effects of water and feed restriction on body weight change and nitrogen balance in desert goats fed high and low quality forages. Small Ruminant Research 2001;41(1; 1) 19-27.

[50] Asplund JM, Pfander WH. Effects of water restriction on nutrient digestibility in sheep receiving fixed water: feed ratios. Journal of Animal Science 1992;6: 1271-1274.

[51] Musimba NKR, Pieper RD, Wallace JD, Galyean ML. Influence of watering frequency on forage consumption and steer performance in Southeastern Kenya. Journal of Range Management 1987;40(5) 412-415.

[52] Ajibola A. The effect of water deprivation and atropine administration on gastro-intestinal function in goats. M.Sc thesis. University of Pretoria; 2000.

[53] Burgos MS, Senn M, Sutter F, Kreuzer M, Langhans W. Effect of water restriction on feeding and metabolism in dairy cows. *American Journal of Physiology: Regulatory, Integrative and Comparative Physiology* 2001;280: R418–R427.

[54] Misra AK, Singh K. Effect of water deprivation on dry matter intake, nutrient utilization and metabolic water production in goats under semi-arid zone of India. Small Ruminant Research 2002;46(2;2) 159-165.

[55] Gómez-Pastén M, Mora O, Pedraza-Chaverri, Shimida A. The effect of a long term feed restriction on metabolism and tissue composition of goats. The Journal of Agricultural Science 1999;132: 227-232.

[56] Pulina G, Bomboi G, Mazzette FB, Dimauro C, Rassu SPG, Nudda A. Changes in metabolic and endocrine measurements during feed restriction in dairy ewes with different BCS. Journal of Animal Science 2007;85(supplement 1) 661.

[57] Hooda OK, Naqvi SMK. Effect of thermal load and feed restriction on relative adaptability of Malpura and Avikalin sheep in semi-arid region. Indian Journal of Animal Science 1990;60: 608–611.

[58] Rhind SM, Mc Neilly AS. Follicle populations, ovulation rates and plasma profiles of LH, FSH and prolactin in Scottish Blackface ewes in high and low levels of body condition. Animal Reproduction Science 1986;10: 105–115.

[59] Maurya VP, Naqvi SMK, Mittal JP. Effect of dietary energy balance on physiological responses and reproductive performance of Malpura sheep in the hot semi-arid regions of India. Small Ruminant Research 2004;55(1-3) 117-122.

[60] Ghanem AM, Barbour EK, Hamadeh SK, Jaber LS, Abi Said M. Physiological and chemical responses in water-deprived Awassi ewes treated with vitamin C. Journal of Arid Environments 2008;72: 141-149.

[61] Ghanem A. The effect of vitamin C supplementation on some physiological an immunological indicators in water-deprived Awassi ewes. Master thesis. American University of Beirut; 2005. p.1-89

[62] Chedid M. Physiological Responses of Feed and Water Restricted Dry Awassi Ewes to Aspirin Administration. Master thesis. American University of Beirut; 2009. p.1-64.

[63] Karnib M. The Effect of Vitamin C Administration on some Physiological Parameters in Water and Feed Restricted Dry Awassi Ewes. Master Thesis, American University of Beirut; 2009. p.1-48

[64] Jaber LS, Hanna N, Barbour EK, Abi Said M, Rawda N, Chedid M, Hamadeh SK. Fat mobilization in water restricted Awassi ewes supplemented with vitamin C. Journal of Arid Environment 2011;75(7) 625-628.

[65] MacFarlane WV, Morris RJH, Howard B, McDonald J, Budtz-Olsen OE. Water and electrolyte changes in tropical Merino sheep exposed to dehydration during summer. Australian Journal Agricultural Research 1961;12(5) 889– 912.

[66] Epstein H. The Awassi sheep with special reference to the improved dairy type. FAO Animal and Health Paper 1985;57: 34-38.

[67] Acharya RM. Small ruminant production in arid and semi-arid Asia. In: Timon VM, Hanrahan JP. (eds). Small ruminant production in the developing countries: Proceedings of an Expert Consultation, 8–12 July 1985, Sofia, Bulgaria. FAO animal production and health paper 1986;58 148-169.

[68] Ermias E, Yami A, Rege JEO. Fat deposition in tropical sheep as adaptive attribute to periodic feed fluctuation. Journal of Animal Breeding and Genetics 2002;119(4) 235-246.

[69] Savendahl L, Underwood LE. Fasting increases serum total cholesterol, LDL cholesterol and apolipoprotein B in healthy, nonobese humans. Journal of Nutrition 1999;129: 2005-2008.

[70] Umunna NN, Chineme CN, Saror DI, Ahmed A, Abed S. Response of Yankasa sheep to various lengths of water deprivation. Journal of Agricultural Science 1981;96: 619-622.

[71] Abdelatif AM, Ahmed MMM. Water restriction, thermoregulation, blood constituents and endocrine responses in Sudanese desert sheep. Journal of Arid Environment 1994;26: 171–180.

[72] Varady KA, Roohk DJ, Loe YC, McEvoy-Hein BK, Hellerstein M.K. Effects of modified alternate-day fasting regimens on adipocyte size, triglyceride metabolism, and plasma adiponectin levels in mice. Journal of Lipid Research 2007;48: 2212-2219.

[73] Dunshea FR, Bell AW, Trigg TE. Relations between plasma nonesterified fatty acid metabolism and body tissue mobilization during chronic undernutrition in goats. British Journal of Nutrition 1988;60: 633–644.

[74] Casamassima D, Pizzo R, Palazzo M, D'alessandro AG, Martemucci G. Effect of water restriction on productive performance and blood parameters in Comisana sheep reared under intensive condition. Small Ruminant Research 2008;78: 169–175.

[75] Annison EF, White RR. Glucose utilization in sheep. Biochemistry Journal 1961;80: 162-169.

[76] Bergman E. N. Production and utilization of metabolites by the alimentary tract as measured in portal and hepatic blood. In: MacDonald I.W., Warner A.C.I (Ed.) Digestion and Metabolism in the Ruminants. University of New England Publishing Unit; 1975. p 292–305.

[77] McDowell GH. Hormonal control of glucose homoeostasis in ruminants. In: Westerterp K.R. (ed): Proceedings of the Nutrition Society 1983;42(2) 149-167.

[78] Bassett JM. Dietary and gastro-intestinal control of hormones regulating carbohydrate metabolism in ruminants. In: MacDonald I.W., Warner A.C.I (Ed.) Digestion and Metabolism in the Ruminants. University of New England Publishing Unit; 1975. p383–398.

[79] Vernon RG. Effects of diet on lypolysis and its regulation. In: Westerterp K.R. (ed): Proceedings of the Nutrition Society 1992;51(3) 397-408.

[80] Houseknecht KL, Portocarrero CP. Leptin and its receptors: Regulators of whole-body energy homeostasis. Domestic Animal Endocrinology 1988;15: 457475.

[81] Chilliard Y, Delavaud C, Bonnet M. Leptin expression in ruminants: Nutritional and physiological regulations in relation with energy metabolism. Domestic Animal Endocrinology 2005;29(1) 2-33.

[82] Schaefer A L, Jones SD M, Tong AKW, Lepage P, Murray NL. The effects of withholding feed and water on selective blood metabolites in market-weight beef steers. Canadian Journal of Animal Science 1990;70: 1151–1158.

[83] Dahlborn K, Holtenius K, Olsson K. Effects of intraruminal loads of saline or water followed by voluntary drinking in the dehydrated lactating goat. Acta Physiologica Scandinavica 1988;132: 67–73.

[84] Olsson K, Dahlborn K. Fluid balance during heat stress in lactating goats. Quarterly Journal of Experimental Physiology 1989;(74) 645–659.

[85] Li B T, Christopherson R J, Cosgrove S.J. Effect of water restriction and environmental temperatures on metabolic rate and physiological parameters in sheep. Canadian Journal of Animal Science 2000;80: 97–104.

[86] Sneddon JC. Physiological effects of hypertonic dehydration on body fluid pools in arid-adapted mammals. How do Arab-based horses compare. Comparative Biochemistry and Physiology 1993;104(A) 201–213.

[87] Ashour G, Benlamlih S. Adaptation of Mediterranean breeds to heat stress and water deprivation. In: Guessous F, Rihani N, Ilham A. (eds.), Livestock Production and Climatic Uncertainty in the Mediterranean: Proceedings of the Joint ANPA-EAAPCI-HEAM-FAO Symposium. Wageningen Pers, Wageningen; 2001.

[88] Caldeira A, Belo C, Santos M, Vazques A, Portugal AV. The effect of body condition score on blood metabolites and hormonal profiles in ewes. Small Ruminant Research 2007(a);68(3) 233–241.

[89] Caldeira A, Belo C, Santos M, Vazques A, Portugal AV. The effect of long-term feed restriction and over-nutrition on body condition score, blood metabolites and hormonal profiles in ewes. Small Ruminant Research 2007(b);68(3) 242–255.

[90] Sykes AR. An assessment of the value of plasma urea nitrogen and albumin concentrations as monitors of the protein status of sheep. British Society of Animal Production Occasional Publication 1978;1: 143-154.

[91] Alamer M. Effect of deprivation and season on some biochemical constituents of blood in Awassi and Najdi sheep breeds in Saudi Arabia. Journal of Animal and Veterinary Advances 2005;48: 15–20.

[92] Hamadeh SK, Hanna N, Barbour EK, Abi Said M, Rawda N, Chedid M, Jaber LS. Changes in physiological and blood parameters in water restricted Awassi ewes supplemented with different levels of Vitamin C. In: European Federation for Animal Science EAAP, 60th Annual Meeting, Barcelona 27 August 2009. Session S.26 Abstract no. 3175. http://www.eaap.org/Previous_Annual_Meetings/2009Barcelona/Papers/26_Hamadeh.pdf

[93] Cork SC, Halliwell RW. The veterinary laboratory and field manual. Nottingham University Press; 2002.

[94] El-Sherif MMA, Assad F. Changes in some blood constituents in Barki ewes during pregnancy and lactation under semi-arid conditions. Small Ruminant Research 2001;40: 269-277.

[95] Moorby JM, Dewhurst RJ, Evans RT, Fisher WJ. Effects of level of concentrate feeding during the second gestation of Holstein–Friesian dairy cows. 2. Nitrogen balance and plasma metabolites. Journal of Dairy Science 2002;(85) 178–189.

[96] Huntington GB, Archibeque SI. Practical aspects of urea and ammonia metabolism in ruminants. Journal of Animal Science 2000;77(E-Supplement) 1-11.

[97] Moen RA, DelGuidice GD. Simulating nitrogen metabolism and urinary urea nitrogen: Creatinine ratios in ruminants. Journal of Wildlife Management 1997;61: 881–894.

[98] Kataria N, Kataria AK. Compartmental water management of Marwari sheep. Veterinarski arhiv 2007; 77(6) 551-559.

[99] Keenan D M, Allardyce C J. Changes of plasma creatinine levels of sheep during submaintenance feeding. Australian Veterinary Journal 1986;63: 29–30.

[100] Olsson K, Benlamlih S, Hossaini-Hilali J, Dahlborn K. Regulation of fluid balance in goats and sheep from dry areas. Options Mediterranennes CIHEAM-IAMZ. 1997;34: 159- 171.

[101] Li KE, Mousa HM, Hume JD. Total body water and water economy in camels, desert goats and desert sheep during water restriction and deprivation. Iugoslavica Physiologica et Pharmacologica Acta 1982; 18: 229- 236.

[102] Marini JC, Klein JD, Sands JM, Van Amburgh ME. Effect of nitrogen intake on nitrogen recycling and urea transporter abundance in lambs. Journal of Animal Science 2004;82(4) 1157-1164.

[103] Jacob RH, Pethick DW, Clark P, D'Souza DN, Hopkins DL, White J. Quantifying the hydration status of lambs in relation to carcass characteristics. Australian Journal of Experimental Agriculture 2006;46(4) 429–437.

[104] Qinisa MM, Boomker EA, Mokoboki HK. Physiological Responses of Water-Restricted Tswana and Boer Goats. Life Science Journal 2011;8(S2) 106-111.

[105] Rawda N. The effect of water restriction on some physiological and immunological indicators in dry and lactating Awassi ewes. Master thesis. American University of Beirut; 2003. p.1-123.

[106] Hanna N. Physiological Changes in Water Stressed Awassi Ewes Supplemented with Vitamin C. Master Thesis. American University of Beirut; 2006. p.1-85.

[107] McKinley MJ, Evered MD, Mathai ML. Renal Na excretion in dehydrated and rehydrated adrenalectomized sheep maintained with aldosterone. American Journal of Physiology: Regulatory, Integrative and Comparative Physiology 2000;279: 17-24.

[108] Maltz E, Olsson K, Glick SM, Fyhrquist F, Shanikout N, Chosniak I, Shkolnik A.Homeostatic responses to water deprivation or hemorrhage in lactating and non-lactating Bedouin goats. Comparative Biochemistry and Physiology 1984;A(77A) 79-84.

[109] Hossaini-Hilali J, Benlamlih S, Dahlborn K. Effects of dehydration, rehydration, and hyperhydration in the lactating and non-lactating black Moroccan goat. Comparative Biochemistry and Physiology 1994;109(A) 1017–1026.

[110] Mengistu U, Dahlborn K, Olsson K. Mechanisms of water economy in lactating Ethiopian Somali goats during repeated cycles of intermittent watering. Animal 2007;1: 1009–1017.

[111] Shaham D, Choshniak I, Rosenfeld J, Witenberg C, Thurau K, Shkolnik A. Modulation of plasma arginine vasopressin during rehydration in the Bedouin goat. Journal of Comparative Physiology B: Biochemical, Systemic, and Environmental Physiology 1994;164(2) 112-117.

[112] 112-Yesberg N, Henderson M, Budtz-Olsen OE. The excretion of vasopressin by normal and dehydrated sheep. Australian Journal of Experimental Biology and Medical Science 1970;48: 115–127. http://www.nature.com/icb/journal/v48/n1/abs/icb197011a.html (accessed 16 August 2012)

[113] Olsson K. Fluid balance in ruminants: adaptation to external and internal challenges. Annuals of the New York Academy of Sciences. Trends in Comparative Endocrinology and Neurobiology 2005;104(1) 156-161.

[114] Abdelatif AM, Elsayed SA, Hassan YM. Effect of state of hydration on body weight, blood constituents and urine excretion in Nubian goats (Capra hircus). World Journal of Agricultural Sciences 2010;6(2) 178-188.

[115] Mojabi A, Abbasali Pourkabire M, Safi S, Bokaie S, Shariati T. Measurements of reference values of some biochemical parameters in serum samples of Ghezel breed sheep. Journal of the Faculty of Veterinary Medicine University of Tehran 2000;55(2) 19-27

[116] Eshratkhah B, Sadaghian M, Safari Nezhad M,Sabri V, Farahmand Geyglou B. Evaluation of electrolytes normal values in blood of Moghani sheep breed. Journal of Animal and Veterinary Advances 2008; 7(4) 437-440.

[117] Mengistu UK, Dahlborn K, Olsson K. Effect of intermittent watering on water balance and feed intake in Male Ethiopian Somali goats. Small Ruminant Research 2007;67: 45 – 54.

[118] Tasker J.B. Fluids, electrolytes, and acid–base balance. In: Kaneko J.J., Cornelius C.E. (ed.), Clinical Biochemistry of Domestic Animals, second ed. Academic Press Inc., London; 1971. p65–75.

[119] Srikandakumar A, Johnson EH, Mahgoub O. Effect of heat stress on respiratory rate, rectal temperature and blood chemistry in Omani and Australian Merino sheep. Small Ruminant Research 2003;49: 193–198.

[120] El-Nouty FD, El-Naggar MI, Hassan GA, Salem MH. Effect of lactation on water requirements and metabolism in Egyptian sheep and goats. World Review of Animal Production 1991;XXVI: 40– 43.

[121] Schneider E G. In water deprivation osmolality become an important determination of aldosterone secretions. News in Physiological Sciences 1990;5: 197–201.

[122] Parker AJ, Hamlin GP, Coleman CJ, Fitzpatrick L. Dehydration in stressed ruminants may be the result of a cortisol-induced diuresis. Journal of Animal Science 2003;81: 512– 519.

[123] Parrott RF, Thornton SN. Effects of psychological stress and dehydration on plasma cortisol, oxytocin and vasopressin in Clun Forest sheep. Journal of Physiology 1988;399(Suppl.) 91.

[124] Parrot RF, Lloyd DM, Goode JA. Stress hormone response of sheep to food and water deprivation at high and low ambient temperatures. Animal Welfare 1996;5: 45– 56.

[125] Fell LR, Lynch JJ, Adams DB, Hinch GN, Munro RK, Davies HI. Behavioral and physiological effects in sheep of a chronic stressor and a parasite challenge. Australian Journal of Agricultural Research 1991;42: 1335-1346.

[126] Huszenicza GY, Kulcsar M, Rudas P. Clinical endocrinology of thyroid gland functions in ruminant. Veterinarni Medicina, 2002;47(7): 199-210.

[127] Latimer KS, Mahaffy EA, Prasse KW. Duncan and Prasses, Veterinary laboratory medicine, Clinical pathology. Forth edition, Iowa state press, USA; 2003

[128] Thrall MA. Veterinary hematology and clinical chemistry. Lippincott Williams & Wilkins; 2004.

[129] Hamadeh SK, Moussa Z, Abi Said M, Barbour E. Physiological indicators of adaptation in Awassi and Finn x Texel x Awassi sheep. Options Méditerranéennes 1994;33 : 231-236.

[130] El-Barody MAA, Abdalla EB, Abd El-Hakeam AA. The changes in some blood metabolites associated with the physiological response in sheep. Livestock Production Science 2002;75(1) 45-50.

[131] Novoselec J, Antunović Z, Šperanda M, Steiner Z, Šperanda T. Changes of thyroid hormones concentration in blood of sheep depending on age and reproductive status. Italian Journal Animal Science 2009;8(3) 208-210.

[132] Ward MA, Neville TL, Reed JJ, Taylor JB, Hallford DM, Soto-Navarro SA, Vonnahme KA, Redmer DA, Reynolds LP, Caton JS. Effects of selenium supply and dietary restriction on maternal and fetal metabolic hormones in pregnant ewe lambs. Journal of Animal Science 2008;86: 1254–1262.

[133] Nazifi S, Gheisari HR, Poorabbas H. The influences of thermal stress on serum biochemical parameters of dromedary camels and their correlation with thyroid activity. Comparative Haematology International 1999;9(1) 49-53.

[134] Yokus B, Cakir DU, Kanay Z, Gulten T, Uysal E. Effects of seasonal and physiological variations on the serum chemistry, vitamins and thyroid hormone concentrations in sheep. Journal of Veterinary Medicine 2006;A (53) 271–276.

[135] Nazifi S, Saeb M, Rowghani E, Kaveh K. The influences of thermal stress on serum biochemical parameters of Iranian fat-tailed sheep and their correlation with triiodothyronine (T 3), thyroxine (T4) and cortisol concentrations. Comparative Clinical Pathology 2003;12(3) 135-139.

[136] More T, Sahni KL. Recent observations on water economy and sheep production under semi-arid conditions. Indian Society of Desert Technology and University Centre of Desert Studies, Transactions 1980;5(1) 115-124.

[137] Olsson K, Benlamlih L, Dahlborn K, Fyhrquist F. Effects of water deprivation and hyperhydration in pregnant and lactating goats. Acta Physiologica Scandinavica 1982;115(3) 361–367.

[138] Rodriguez MN, Tebot I, Le Bas A, Nievas C, Leng L, Cirio A. Renal functions and urea handling in pregnant and lactating Corriedale ewes. Canadian Journal of Animal Science 1996;76: 469–472.

[139] Sibanda LM, Ndlovu LR, Bryant MJ. Effects of feeding varying amounts of a grain/ forage diet during late gestation and lactation on the performance of Matebele goats. The Journal of Agricultural Science 1997;128(4) 469-477.

[140] Mittal JP, Ghosh PK. Effect of prolonged intermittent water restriction on the reproductive performance of ewes in the Indian desert. Animal Production 1986;43(2)255-260.

[141] Mellado M, Valdéz R, García JE, López R, Rodríguez A. Factors affecting the reproductive performance of goats under intensive conditions in a hot arid environment. Small Ruminant Research 2006;63: 110-8.

[142] Laporte-Broux B, Roussel S, Ponter AA, Perault J, Chavatte-Palmer P, Duvaux-Ponter C. Journal of Animal Science 2011;89: 2154-2163.

[143] Sevi A, Albenzio M, Annicchiarco G, Caroprese M, Marino R, Taibi L. Effects of ventilation regimen on the welfare and performance of lactating ewes in summer. Journal of Animal Science 2002;8: 2362– 2372.

[144] Lindsay DB, Leat, WMF. Carbohydrate and lipid metabolism. In: Hunt M.H. (ed.) The Blood of Sheep Composition and Function. Springer Verlag, Berlin, Heidelberg, New York; 1975. p. 53.

[145] Fell BF, Mackie WS, Campbell RM. Quantitative changes occurring in the serum proteins during lactation in the ewe. Research in Veterinary Science 1968;9: 563–572.

[146] Collier RJ. Nutritional, metabolic, and environmental aspects of lactation. In: Larson B.L. (ed.). Lactation. Iowa State University Press 1985.p. 102–107.

[147] Maltz E, Shkolnik A. Milk production in the desert: lactation and water economy in the black Bedouin goat. Physiological Zoology 1980;53: 12-18.

[148] Dahlborn K, Nielsen MO, Hossaini-Hilali J. Mechanisms causing decreased milk production in water deprived goats. Options Mediterranéennes CIHEAM- Zaragoza 1997;34: 199-202.

[149] Dahlborn K. Effect of temporary food or water deprivation on milk secretion and milk composition in the goat. Journal of Dairy Research 1987;54(2) 153-163.

[150] Alamer M. Effect of water restriction on lactation performance of Aardi goats under heat stress conditions. Small Ruminant Research 2009;84(1) 76-81.

[151] Mittal SB. Dairy development and seasonal variation in the quality of milk in the arid zone of western Rajasthan. In: H. S. Mann (ed.) Arid zone research and development. Scientific Publishers, Jodhpur, India; 1980. p381-387.

[152] Yagil R, Amir H, Abu-Rabiya Y, Etzion Z. Dilution of milk: a physiological adaptation of mammals to water stress? Journal of Arid Environments 1986;11: 243-247.

[153] Renaudeau D, Collin A, Yahav S, de Basilio V, Gourdine JL, Collier RJ. Adaptation to hot climate and strategies to alleviate heat stress in livestock production. Animal 2012;65: 707–728.

[154] Apple JK, Minton JE, Parsons KM, Unruh JA. Influence of repeated restraint and isolation stress and electrolyte administration on pituitary-adrenal secretions, electrolytes, and other blood constituents of sheep. Journal of Animal Science 1993;71: 71-77.

[155] Baldock NM, Sibly RM. Effects of handling and transportation on the heart rate and behaviour of sheep. Applied Animal Behaviour Science 1999;28(1–2) 15-39.

[156] Ali BH, Al-Qarawi AA, Mousa HM. Stress associated with road transportation in desert sheep and goats, and the effect of pretreatment with xylazine or sodium betaine. Research in Veterinary Science 2005;80 (3) 343-348.

[157] Schaefer AL, Jones SDM, Stanley RW. The use of electrolyte solutions for reducing transport stress. Journal of Animal Science 1997;75: 258–265.

[158] Al-Haidary AA. Physiological responses of naimey sheep to heat stress challenge under semi-arid environments. International Journal of Agriculture and Biology 2006;6(2) 307–309.

[159] Christensen RA, Christopherson RJ, Kennedy JJ. Effect of somatostatin and chronic cold exposure on hormonal and metabolite concentration, metabolic rate, thermoregulation and gut motility in sheep. Canadian Journal of Animal Science 1990;70: 1073–1083.

[160] Abdel-Samee AM, Abd-Alla OAM, EL-Adawy SAI. Nutritional treatments for alleviation of heat stress in Awassi sheep using acacia and olive pulp in subtropics. Egypt. J. Comp. Path. & Clinic. Path. 2008;21(1) 466- 477.

[161] Morrison SR. Ruminant heat stress: effect on production and means of alleviation. Journal of Animal Science 1983;57(6) 1594-1600.

[162] Darcan N, Güney O. Alleviation of climatic stress of dairy goats in Mediterranean climate. Small Ruminant Research 2008;74: 212–215.

[163] Ayo JO, Minka NS, Mamman M. Excitability scores of goats administered ascorbic acid and transported during hot-dry conditions. Journal of Veterinary Science 2006;7(2) 127–131.

[164] Minka NS, AYO JO. Physiological responses of transported goats treated with ascorbic acid during the hot-dry season. Animal Science Journal 2007;78: 164–172.

[165] Hathcock JN, Azzi A, Blumberg J, Bray T, Dickinson A, Frei B, Jialal I, Johnston CS, Kelly FJ, Kraemer K, Packer L, Parthasarathy S, Sies H, Traber MG. Vitamins E and C are safe across a broad range of intakes. American Journal of Clinical Nutrition 2005;81(4) 736-745.

[166] Xu Z, Kornegay ET, Sweet LA, Lindmann MD, Veit HP, Watkins BA. Effects of feeding aspirin and soybean oil to weanling pigs. Journal of Animal Science 1990;68: 1639-1647.

[167] Fields M, Lewis C, Bureau I. Aspirin reduces blood cholesterol in copper deficient rats: A potential antioxidant agent? Metabolism 2001;50(5) 558-561.

[168] Drew J, Arthur J, Farquharson A, Russel W, Morrice P, Duthie G. Salicylic acid modulates oxidative stress and glutathione peroxidase activity in the rat colon. Biochemical Pharmacology 2005;70(6) 888-893.

[169] McDowell LR. Vitamins in animal and human nutrition. State University, Ames IA, Iowa; 2000.

[170] de Rodas BZ, Maxwell CV, Davis ME, Mandali S, Broekman E, Stoecker BJ,. L-ascorbyl-2-polyphosphate as a vitamin C source for segregated and conventially weaned pigs. Journal of Animal Science 1998;76: 1636–1643.

[171] Avci M, Yertürk M, Kaplan O. Effects of ascorbic acid on the performance and some blood parameters of Japanese quails reared under hot climate conditions. Turkish Journal of Veterinary and Animal Sciences 2005;29: 829-833.

[172] Yousef MI. Aluminium-induced changes in hemato-biochemical parameters, lipid peroxidation and enzyme activities of male rabbits: protective role of ascorbic acid. Toxicology 2004; 47-57.

[173] McKee JS, Harrison PC, Riskowski GL. Effects of supplemental ascorbic acid on the energy conversion of broiler chicks during heat stress and feed withdrawal. Poultry Science 1997;76: 1278-1286.

[174] Jariwalla R. J. ,Harakech S. In: Harris, R.J. (ed.) Ascorbic acid: Biochemistry and biomedical cell biology, Plenum, New York; 1996. p. 215-231.

[175] Bouda J, Jagos P, Dvorak R, Ondrova J. Vitamin E and C in the blood plasma of cows and their calves fed from buckets. Acta Veterinaria Brno 1980;49: 53-58.

[176] Politis I, Hidiroglou M, Batra TR, Gilmore JR, Gorewit RC, Scherf H. Effects of vitamin E on immune function of dairy cows. American Journal of Veterinary Research 1995;56: 179.

[177] Cummins KA, Brunner CJ. Effect of calf housing on plasma ascorbate and endocrine and immune function. Journal of Dairy Science 1990;74: 1582-1588.

[178] Sahin K, Kucuk O, Sahin N, Sari M. Effects of vitamin C and vitamin E on lipid peroxidation status, serum hormone, metabolite, and mineral concentrations of Japanese quails reared under heat stress (34 1C). International Journal of Vitamin Nutrition Research 2002;72: 91–100.

[179] Kannan G, Terrill H, Kouakou B, Gazal OS, Gelaye S, Amoah EA, Samake S. Transportation of goats: Effects on physiological stress responses and liveweight loss. Journal of Animal Science. 2000;78: 1450–1457.

[180] Mahan DC, Ching S, Dabrowski K, Developmental aspects and factors influencing the synthesis and status of ascorbic acid in the pig. Annual Review of Nutrition 2004;24: 79–103.

[181] Civen M, Leeb JE, Wishnow RM, Morin R. Effect of dietary ascorbic acid and vitamin E deficiency on rat adrenal cholesterol ester metabolism and corticosteroidogenesis. International Journal of Vitamin Nutrition Research 1980;50: 70–78.

[182] Sivakumar AVN, Singh G, Varshney VP. Antioxidants supplementation on acid base balance during heat stress in goats. Asian-Australian Journal of Animal Science 2010;23(11) 1462 – 1468.

[183] Jeffreys D. Aspirin: The story of a wonder drug. London: Bloomsbury Publishing; 2004.

[184] He J, Whelton PK, Klag MJ. Aspirin and risk of hemorrhagic stroke: a meta-analysis of randomized controlled trials. The Journal of the American Medical Association 1998;280: 1930-1935.

[185] Galil MA. Effect of using some anti-heat stress compounds on the performance of some local breeds of chicken under hot climatic condition. Egyptian Poultry Science Journal 2004;24(2) 417-427.

[186] McDaniel CD, Balog JM, Freed M, Elkin RG, Wellenreiter RH, Hester PY. Response of layer breeders to dietary acetylsalicylic acid. 1. Effects on Hen Performance and Eggshell Quality. Poultry Science 1993;72(6) 1084-1092

[187] Stilborn HL, Harris GC Jr, Bottje WG, Waldroup PW. Ascorbic acid and acetylsalicylic acid (aspirin) in the diet of broilers maintained under heat stress conditions. Poultry Science 1988;67(8) 1183-1187.

[188] Traş B, Inal F, Baş AL, Altunok V, Elmas M, Yazar E. Effects of continuous supplementations of ascorbic acid, aspirin, vitamin E and selenium on some haematological parameters and serum superoxide dismutase level in broiler chickens. British Poultry Science 2000;41: 664-66.

[189] Gudev D, Popova-Ralcheva S, Yanchev I, Kozelov L, Moneva P. Effects of acetylsalicylic acid on adrenal response in lambs exposed to transport and new environment. Bulgarian Journal of Agricultural Science 2002;8(4) 433-438.

[190] Pollard JC, Roos V, Littlejohn RP. Effects of an oral dose of acetyl salicylate at tail docking on the behaviour of lambs aged three to six weeks. Applied Animal Behaviour Science 2000;71: 29-42.

Chapter 7

Water Stress and Agriculture

Sonia Marli Zingaretti, Marielle Cascaes Inácio, Lívia de Matos Pereira, Tiago Antunes Paz and Suzelei de Castro França

Additional information is available at the end of the chapter

http://dx.doi.org/10.5772/53877

1. Introduction

Climate changes and water availability cause an important impact in agriculture, food disposal and consequently in human health. According to the U.S. Census Bureau, the total population of the World is now over 7,032 billion, and all growth projections for developed and undeveloped countries show that a total of 9 million of inhabitants will be reached before 2050. As a result, the demand for food and fuel will increase significantly. How agriculture will move on to guarantee continuous provision of food for all inhabitants of the planet?

As a consequence of the population growth the scenario has changed and an increase of urban areas has occurred in the last decades. From the $50^{th's}$ urban population increased from 4 to 11% in Africa, 33 to 52% in Asia, 9 to13% in Latin America, and decreased from 38 to 15% in Europe, from 15 to 6% in Northern America. Growth forecasts for 2050 are 54, 32.5 and 6.8% increase in Asia, Africa and Latin America respectively. Increasing urban areas make less cropland available for fuel and food production. Croplands are not expanding in the same rate as population in the last half a century and salinity and desertification have also contributed to the fact that less useful areas remain proper for agriculture. According to FAO about 12% of globe´s land surface is used for crop production and most of remaining world agricultural land are covered by forest and protect by environmental laws. Brazil, Bolivia, Argentina, Colombia, Sudan and Democratic Republic of the Congo retain 90% of accessible agricultural land [1].

The climate changes in agriculture and human life can be considered under different aspects: the biological effects on crop yields; the resulting impacts on outcomes including prices, production and consumption and also the impact on per capita calorie consumption and child malnutrition [2]. In other words their effects on agriculture will induce changes in production and prices, altering economic system, crop mix, production, food demand and consumption. Unfortunately those changes are already occurring and the projections on annual mean temperature for the next 20 to 30 years point to great economic losses due to decline in productivity for cereals like maize, wheat and rice as well. It is well known that most of our important crops will decrease yield with temperature above 30°C, as they growth faster in high temperature they have less time to accumulate carbohydrates, proteins and oil. Increasing temperature will perhaps make some areas available for agriculture, but will it be enough to replace the areas that will certainly be lost?

Recently, it was discussed the physical and economic consequences of climate changes considering temperature rising in Europe over four different factors such as agriculture, river floods, coastal systems and tourism [3]. Considering four different temperature increases from 2.5 to 5.1°C and five Europe regions (Southern, Central South, Central North, British Isle and Northern). Yield change (%) would affect Southern Europe (Portugal, Spain, Italy, Greece and Bulgaria) more than any other region with temperature increase. Northern Europe (Sweden, Finland, Estonia, Latvia, and Lithuania) instead would benefit from positive yield changes. River floods are natural disasters anywhere it happens, resulting in very large economic losses due to properties and agriculture damage. An increase on river flood is expected with global warming [4]. As a consequence of increasing temperature river floods would affect 250,000-400,000 additional people in Europe in the 2080s, specially Western Europe, British Isle and Central Europe regions. All the costal systems across Europe would suffer with people flooded. Tourism in Europe would be impacted as well. According to bed night's percentage measures the effects will be a decrease in Southern Europe and an increase in all other areas such as Central and Northern Europe. But not only temperature would have importance to agriculture, fluctuation in seasonal precipitation is also extremely relevant and as well as increasing evaporation rates [3].

The effects of climate change on rainfed and irrigated crops for developing and developed countries were also discussed [2]. Percentage change in yield for irrigated and rainfed crops like maize, rice and wheat were analyzed using Decision Support System for Agrotechnology Transfer (DSSAT) crop-simulation model with and without CO_2 fertilization in 2050 scenario. The observed effects on rainfed were attributed to changes in temperature and precipitation index, while for irrigated areas the effects were only related to temperature variation. In general, yields in developed countries were less affected than those in developing countries, where for most crops without CO_2 fertilization the yield declines. The stress imposed by climate changes on agriculture will certainly intensify the disparities among regions.

Nevertheless, prices for major grain crops like rice, wheat, maize and soybean will increase up to 60 to 70 %, over the next few years, even without climate changes. Bearing in mind the predicted weather changes an additional of 32 to 37% for rice, 52 to 55% for maize, 94 to 111% for wheat and 11 to 14% for soybean can be expected [2].

2. Physiological aspects of water stress

It is well known that plant growth and development can be affected by abiotic agents such as salinity, high temperatures, radiation, flood and water deficit. Exacerbate action of those environmental conditions can led to great losses in productivity due to crop stress. When subjected to water deficit plants go through a cascade of metabolic alterations started with reduction in photosynthetic pigments concentration. Physiological mechanisms of plant response to water stress are summarized in Figure 1. Facing a water deficit situation plant responses can be species/genotype specific, under rehydration after a mild water deficit almost every plant can return to normal growth, but if the stress intensity was moderated or severe some will not recover at all.

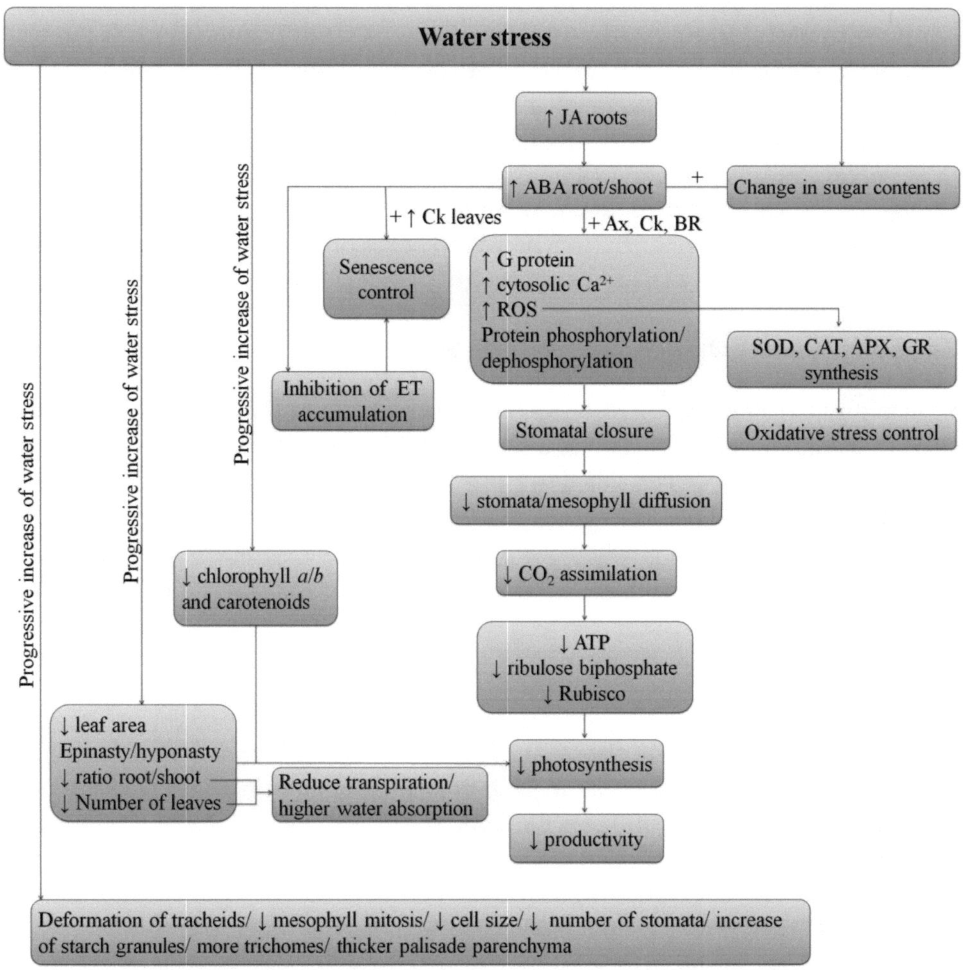

Figure 1. Physiological mechanisms induced by water stress.

2.1. Photosynthetic responses

One of the significant alterations responsible for reduction in crop productivity is low photosynthetic ability. The water stress may cause decrease in CO_2 assimilation in the leaves, the amount of ATP and the level of ribulose bisphosphate [5-8]. Stomatal closure limiting diffusion through stomata and mesophyll is one of the first events in plants response under water deficit situation with consequent increase of the Rubisco enzyme, responsible for CO_2 fixation, e in order to overcome the low conductance [9-13]. However, some species (*Gossypium barbadense, Hypericum balearicum*) show, instead, a decrease in Rubisco activity [13-15]. Furthermore the decrease in CO_2 concentration will induce a reduction in the dynamic of the carboxylation process [8].

The electron transport in thylakoids and the use of trioses phosphates are also reduced in the stress biochemical control therefore, the net photosynthetic rate tends to be lower. It has been proved that in plants subjected to water stress the photochemical efficiency of photosystem II (PSII) and quantum generation is reduced [6, 15-17]. Alterations in the level of photosynthetic pigment were also detected in water stressed plants; showing reduced or even no pigmentation. Both chlorophyll *a* and *b* declines in stressed plants and this directly affect plant biomass production. The reduction of photosynthetic pigments will drive a cut down in energy consumption and carbon demand for chlorophyll synthesis [8, 11, 17, 18]. Other pigments such as carotenoids, which play essential role in the antioxidant defense system under stress conditions, can function as an accessory pigment for photosynthesis although, their concentration can also be reduced as part of plant response [8, 17, 19, 20].

2.2. Sugar and Reactive Oxygen Species (ROS) protection

Changes in the content of carbohydrates such as sucrose and raffinose, together with the unbalanced of plant hormones function as a signal that plant response to stress should be initiated [21]. Raffinose has been correlated to a plant tolerance to desiccation, possibly involved in the protection against ROS that are responsible for loss of membrane integrity and cellular death [22, 23]. Moreover, induction of sugar accumulation, i. e, sucrose, fructose, maltose and inositol is relevant for the osmoprotection process and has been associated to plant tolerance to water stress [22].

2.3. Hormonal regulation

It is well known that hormones play a special role in plant reaction to water stress conditions. The abscisic acid (ABA) is the main hormone correlated to water stress. Plants exposed to drought substantially increase the level of ABA in shoots and roots [24-26], and the ABA positive regulators induce plant response as G protein activation, ROS production, increase in cytosolic Ca^{2+}, protein phosphorylation and dephosphorylation events and immediate stomatal closure [27, 28]. Actually the balance of positive and negative ABA regulators actions command the resistance or sensitivity to water scarcity situation. However, the regulation of stomatal closure occurs not only due the action of ABA, but by the integrated hormonal balance between ABA, Auxin (Ax) and Cytokinin (Ck) [21]. Along with Ck, ABA

http://dx.doi.org/10.5772/53877

plays a role in controlling senescence [29, 30]. The high concentration of ABA possibly prevents excessive accumulation of ethylene (ET), thus indirectly maintaining the growth of roots and shoots [31, 32].

ABA seems to be also involved in remobilize carbon to accelerate grains filling in rice and wheat [29, 30, 33]. The ABA increased level also induce ROS production and in order to prevent the oxidative stress, antioxidant enzymes such as superoxide dismutase (SOD), catalase (CAT), ascorbate peroxidase (APX) and glutathione reductase (GR) are immediately activated [34, 35]. The importance of regulating ABA contents as a stress signaling was also observed in rice leaves when, after a water stress period, plants were rehydrated; decreasing ABA content occurred a reduction on ROS, CAT and SOD [36].

Wheat and maize plants submitted to a moderate water deficit exhibit an increase in ET concentration [27, 37] that would be partly responsible for decreasing growth rates. In contrast for beans, cotton and miniature rose it has been shown that the rate of ethylene production is not affected during progressive drought [27]. ET is also involved in ROS production and antioxidant enzymes synthesis [23].

Cytokinins are generally involved in root and shoot development, but it has been shown an increased cytokinin concentration in leaves, from roots translocation, in plants submitted to water deficit [24, 29, 38]. Bentgrass transgenic plants expressing the enzyme adenine isopentenyl phosphotransferase for Ck synthesis ligated to a senescence-activated promoter (*SAG12*), resulted in increases in Ck accumulation in the leaves and roots and in the overall plant tolerance to water stress [39].

The gibberellins (GA), Ax and brassinosteroids (BR) do not seem to have a direct involvement with water stress, however, the accumulation of GA in some dicots has been reported and also the BR along with ABA regulate the development and function of stomata [27]. In contrast in monocots such as maize, there is a decrease in the levels of GA in leaves [25, 27]. The Ax content in plants seems to decrease in roots and leaves under stress, but the importance of auxin in water stress response remains inconclusive [24, 25, 27, 40].

The JA instead seems to play a role in the biosynthesis of ABA in water stress. In citrus plants, for example, drought causes an increase of JA concentration in roots with subsequent increase in the concentration of ABA. One can conclude that JA is possibly the precursor in the signal transduction cascade in case of drought stress, providing increased levels of ABA which, in turn, induce later responses [41].

2.4. Morphological and anatomical modifications

All the known morphological changes that occur in plants under water deficit can be associated to hormone actions. Plants develop more roots in order to access more water, increasing the ratio root/shoot, reduce leaf number and leaf area to lower transpiration rates what leads, unfortunately, to a decrease in photosynthetic rates and biomass production [20, 26, 42] and develop the epinasty/hyponasty effects [43]. The increase of apoplastic pH in the elongation of leaf area could be the responsible for the foliar reduction [44]. The number of lateral seedling roots [17] as well as the stem length can also be affected [20].

The deformation of tracheids in the xylem due to the decrease in osmotic potential [45], the reduction of mitotic activity of mesophyll cells [46], the increase of starch granules in chloroplasts [17, 47] and trichome production, as well as the decrease in cell size and number of stomata per leaf, the thickness of palisade parenchyma are anatomical changes resulting from water stress [7].

2.5. C_3 and C_4 responses

The response to water deficit may vary from species C_3 i.e. *Triticum aestivum* and *Panicum bisulcatum* and C_4 i.e. *Zuloagaea bulbosa* and *Zea mays*. Although in the oxidative stress the production of hydrogen peroxide seems to be higher in C_3 plants, the C_4 plants are much more sensitive to water stress mainly due to stomatal closure and non-stomatal factors such as decreased activity of photosynthesis enzymes, the induction of premature senescence and changes in leaf anatomy [48-50].

3. Plant biochemical mechanisms to face water stress

Drought stress and its detrimental effects on plants in both natural and agricultural environments are receiving increasing attention in order to discover alternative solutions to enhance plant vigor and high tolerance; to maintaining crop yields under adverse or extreme climate conditions overcoming economic losses.

3.1. Drought effects: Two sides of the coin

Contrasting with the negative environmental aspects caused by water stress, the adverse effects on agriculture affecting plant growth and crop productivity can be mitigated by metabolic changes which invigorate the plant biosynthesis of natural products with widespread use by the pharmaceutical, energy and food industries.

3.2. Protective role of secondary metabolites in the plant response and tolerance to water stress

Plant defense response and tolerance to drought and salinity involves the perception of signal stress by receptors at the membrane level followed by signaling transduction in the cell, inducting a multiplicity of biochemical mechanisms involved in the protective role of secondary metabolites. Water stress reduces plant growth, so the carbon fixed during photosynthesis could be used to form secondary metabolites as established in several studies. Restrictions of water supply to plant bring about the production of a complex variety of secondary metabolites which level can be modulated through biochemical and genetic manipulation. Water stress induce the accumulation of reactive oxygen species (ROS), resulting in oxidative stress in the plant cells. Thus, antioxidant secondary metabolites, able to scavenger and detoxify ROS by the availability of –OH, –NH_2, and –SH groupings, as well as aromatic nuclei and unsaturated aliphatic chains, can play important role in protecting plant species against oxidative stress caused by water deficit. It is well known that terpenoids possess antioxidative properties. Volatile isoprenoids accumulated in *Hevea brasiliensis* were thought to be involved in scavenging ROS and potentially in protecting this species against oxidative stress [51].

Similarly, drought stress markedly enhanced the total concentrations of monoterpenes and resin acids in the main stem wood of Scots Pine and Norway Spruce Seedlings [52]. Results of investigations conducted on the effect of water deficit imposed to potted *Prunella vulgaris* L. a Chinese plant of medicinal and industrial importance, demonstrated increased levels of the phenolic triterpenes rosmarinic acid, ursolic acid and oleanolic acid [53]. Drought also caused increased accumulation of phenolic compounds (ferulic acid) in the leaves of triticale seedlings [54] and enhancement of total phenolic contents of *Trachyspermum ammi* leaves [55]. Investigation conducted with water-stressed cucumber (*Cucumis sativus* L.) demonstrated that adverse effects of water stress can be minimized by the application of melatonin which promotes activities of the ROS scavenging enzymes, i.e., superoxide dismutase, peroxidase, and catalase; significantly reduced chlorophyll degradation and stimulates root generation and vitality [17].

Polyamines (PAs) are low molecular weight polycations that have been implicated in a wide range of biological processes in plant growth and development, including environmental stress. The major PAs occurring in plant cells are the diamine putrescine (PUT), triamine spermidine (SPD) and tetramine spermine (SPM). Among the important roles attributed to those plant polyamines are: membrane stabilization and free radicals scavenger action. Polyamine mediated regulation of the water deficit stress response of soybean seedlings was investigated using exogenous applications of polyamines and their biosynthetic inhibitors. The exogenous supply of PUT, SPD and SPM to soybean seedlings resulted in reduction of the stress injury in roots which showed increased length and water content over non-treated stressed controls. Moreover, up to 40% increase of shoot growth was observed in seedlings supplemented PUT, SPD and SPM in comparison with controls. In contrast, in the presence of polyamines inhibitors the stress injury intensified, growth was severely inhibited, and water content of roots was significantly decreased. Overall results suggested that polyamines are potentially useful to overcome the detrimental effects of drought [56].

Water stress is also known to increase the secondary metabolite production in a variety of medicinal plants. Increase of hypericin and betulinic acid levels upon *Hypericum brasiliense* Choisy exposure to drought represents an antioxidant response to ROS production [57, 58]. *Artemisia annua* plants tolerate well water deficit treatments showing increased accumulation of artemisinin, a potent anti-parasitic drug, as proved in greenhouse experiments. The authors suggested that artemisinin could be part of *A. annua* chemical system of defense against water deficit [59]. Pharmacologically active terpene indole alkaloids production is stimulated in Apocynaceae species in response to water deficit. Comparing *Catharanthus roseus* drought stress plants with well watered controls it was observed significant enrichment in the antihypertensive drug ajmalicine suggesting that the production of this class of secondary metabolite can be upgraded during drought stress *C. roseus* plants [60].

The main physiological and biochemical known mechanisms triggered by water stressed plants are illustrated in Figure 2.

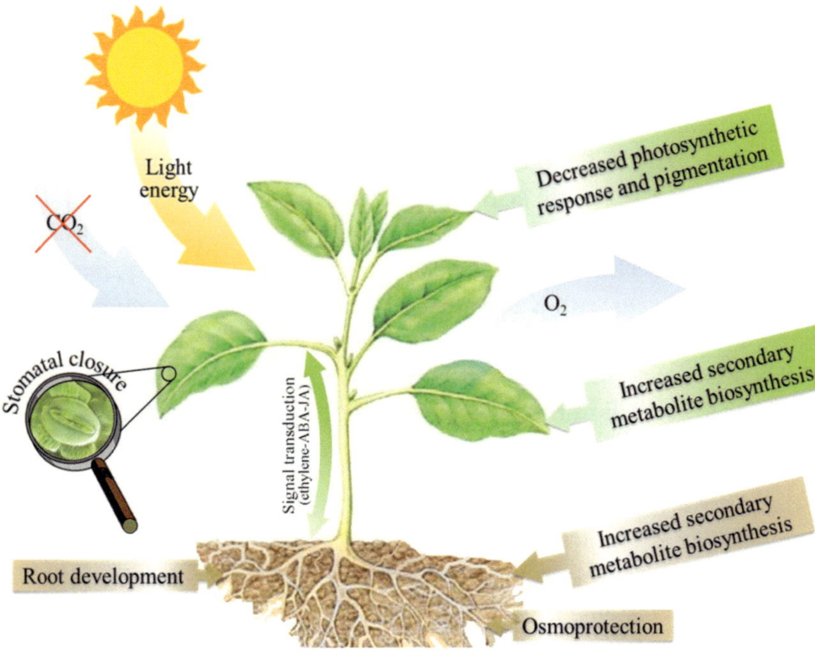

Figure 2. Plant global response to cope with water deficit, high temperature and salinity.

3.3. Sustainable exploitation of cultured drought-resistant plants

3.3.1. Oilseed crops: Biofuel production

Recently, considerable attention has been given to biofuels as an alternative to fossil fuels and the challenge is to find oil bearing plants that produce non-edible oils as the feedstock for biodiesel production. Jatropha (*Jatropha curcas* L.) popularly known in Brazil as "pinhão manso" is a drought resistant perennial tree widespread in arid, semi-arid and tropical regions of the world and requires a minimum rainfall of 250 mm [61]. Native grown in marginal and waste lands is one of the potential plant species to be exploited as a new source of oil for biodiesel production. Jatropha represents great promise to the energy economy of developing, as well as developed countries. It has been reported that new and large markets for biodiesel demand are expected to emerge in China, India and Brazil [62]. In those countries farmers have started to produce Jatropha for biodiesel production. The Jatropha seed is particularly suitable for biodiesel production because it can be harvested in the third year of plantation five or six times annually. Diesel consumption in Brazil is about 40 billion liters per year, providing huge opportunities for biodiesel production, and it is estimated that by 2013 the biodiesel market will be approximately 2 billion liters [63].

Besides the oil production, Jatropha species are source of jatrophone and jatropholone, macrocyclic diterpenoids, secondary metabolites that display varied pharmacological activities [64, 65].

3.3.2. Quinoa

Based on the high quality of the oil, and on the fact that some varieties show oil concentrations of up to 9.5%, quinoa could be considered as a potentially valuable new oil crop [66].

Quinoa is currently grown for its grain in the South American countries of Peru, Bolivia, Ecuador, Argentina, Chile and Colombia. Quinoa populations display a high degree of genetic distancing, and variable tolerance to salinity. Cultivars of quinoa can be adapted to growth from sea level to an altitude of 4,000 m, from 40°S to 2°N latitude, and from the cold highland climate to subtropical conditions, i.e. quinoa plant is cold and drought tolerant. The plasticity of quinoa biochemical response to a wide range of environmental conditions makes it possible to select, adapt, and breed cultivars [67]. Studies have shown that quinoa is a very good source of antioxidants and it can be a substitute for common cereals [68, 69]. The content of total phenolic compounds and the correlated radical scavenging activity of quinoa varieties have been analyzed. There were significant differences between the varieties and the content of total polyphenols [70]. Moreover, the saponins obtained as a by-product in the processing of quinoa grain can be utilized by the cosmetics and pharmaceutical industries.

3.3.3. Cotton

Polyphenols and carotenoids compounds with reactive oxygen species (ROS)-scavenging ability biosynthesized in drought tolerance Cotton genotypes were correlated to the drought tolerance of this important crop [71].

4. What have been done?

In order to cope with the major environmental problems that affect crops such as drought, salinity, cold and heat shock, genetic engineering and breeding techniques have become fundamental tools, as they have been for decades regarding biotic stresses, pests and diseases resistance. It is well known that very often an adversity results in another unfavorable condition for the development of a crop, for example high temperatures provoke water deficit reducing soil moisture resulting in salinity problems and desertification.

Biotechnological approaches focused on secondary metabolism pathways induction or repression at the transcriptional level are now being conducted to significantly improve plant tolerance to water deficit, extreme temperatures and ion imbalance.

4.1. Breeding crops

Considering all climate changes that the planet is going through it is vital the development of crops with high efficiency in water recovery and consequently tolerance to water stress, higher temperatures, salinity and desertification. Through conventional breeding methods and selection based in progeny tests it was possible to obtain stress resistant varieties [72], but it has to be considered that instability of genotypes in different environments may affect the cultivars agronomic performance.

Researchers consider that genetic improvement for stress tolerance can be achieved in two ways: directly, through the evaluation of primary features in the target environment, i.e. as productivity (empirical breeding); or indirectly (analytical breeding), through secondary characteristics related to stress adaptation observed in crops growing in limiting environment. Over the past 50 years genetic improvement have been carried out empirically, however, this type of traditional selection has not presented significant efficiency in terms of productivity, requiring the support of indirect selection [73]. The selection of genotypes with promising agronomic characteristics and tolerant to abiotic stresses demands successive seasonal evaluations of field cultures conducted in different locations, and under influence of stress agents, requesting arduous and extensive work.

Furthermore, it is important to highlight that the low heritability of complex traits have limited the development of tolerant cultivars due to significant G x E interaction and the QTL-by-environment interaction (QTL x E), and the trivial understanding of the physiological parameters related to the genetic yield potential in dry environments [73]. Biotechnology plays an important role for managing abiotic stress, allowing the exploitation of large germplasm collections with no need of experimental procedures under unfavorable environmental conditions [74].

4.2. Which genes are involved in plant responses?

Lately, much has been done to identify and isolate drought-induced genes in order to investigate the role those gene products play and the paths for induction of those genes [75, 76]. Gene expression in response to water stress can enhance the plant's ability to respond appropriately to the deleterious effect of drought, stimulating its aptitude to survive desertification [77]. In general, the stress-induced gene products can be classified in two ways: genes that directly protect the plant against stress and genes that regulate the expression of other genes [78, 79].

Through analysis of transcripts it was observed that the genes exhibit distinct expression profiles, being that stress-induced gene decrease mRNA levels when the plants are freed from stress conditions. However, the expression patterns of those genes are complex, with some genes responding very quickly to water deficit while others answer very slowly after the accumulation of ABA (abscisic acid) [80].

The differential gene expression analysis of two sugar cane cultivars, tolerant and sensitive to drought, showed that the number of genes expressed in the sensitive cultivar increased with the severity of the drought. Comparing the gene expression profiles 91 common genes were found among both cultivars, most of them drought-induced genes that are still unknown. Moreover, genes of important pathways related to drought stress were suppressed in sensitive plants. It was evidenced that plants submitted to the same water conditions responded differently to stress. Morphological changes occurred, but some genes may represent the difference between tolerance and sensitivity, as the S-adenosylmethionein decarboxylase (SAMDC) and induced cinnamoyl-CoA reductase (CCR) in resistant cultivars or lipid transfer protein that have been repressed, as well as other genes [81].

The expression of some sugarcane water-stress related genes and their association with sucrose accumulation was also investigated and a group of stress-induced genes that could be associated with sucrose accumulation were identified, showing that genes associated with the synthesis of proline are associated with stress and sucrose accumulation. Stress-related transcription factors and sugar transporter also play a role in sucrose accumulation [82].

For better understanding the processes and genes involved in water deficit tolerance it is required a full knowledge of the molecular principles that regulate plant responses to stress conditions. Thus, studies with model plants stand for and will continue to represent a relevant strategy for the elucidation of signaling and transcription processes using molecular genetics techniques [79, 83].

Genes isolated from several cultured species have been the focus of researches using gene expression in model plants with the objective of elucidating their direct effect on abiotic stress tolerance. Genetic transformation of plants in order to increase resistance is often based on the manipulation of genes to preserve the function and structure of cellular components [84]. In this context, the genetic engineering techniques for pest and herbicide resistance differ from the procedures for abiotic stress tolerance, since the first is a monogenic trait, more easily manipulated. In contrast, tolerance to environmental stresses may associate more than one of the genes involved in different signaling pathways.

The expression of *SPCP2* that encodes the putative papain-like cysteine protease isolated from senescent leaves of sweet potato has been studied in transgenic Arabidopsis plants subjected to stress conditions and has shown very interesting results. Firstly, changes in phenotypic characteristics were noted, such as alterations in the development of seeds and silique resulting in greater incompatibility and lower production and seed germination. Furthermore, *SPCP2* gene expression caused early transition from vegetative to reproductive stage and foliar senescence, indicating that the gene is associated with senescence. Results also indicated that the gene expression was induced by darkness, ethephon, abiscisc acid (ABA) and jasmonic acid (JA). However, tolerance to salinity and drought stress was increased [85].

4.2.1. Signaling genes

Many of these genes encode proteins involved in signaling pathways, including protein kinases mitogen-activated (MAPKs), histidine kinases, protein kinase Ca^{2+} dependent (CDPKs), family SOS3 sensors Ca^{2+} as well as transcription factors [86, 87].

The association of three genes *SRK2D, SRK2E* and *SRK2I* in ABA signaling and in water stress tolerance, since their gene products were found to be involved in ABRE-protein phosphorylation (ABA responsive element) and ABA signaling during germination and root development and at stomatal level was reported [88]. Genetic transformation and genetic crosses carried out to obtain double and triple mutants of Arabidopsis plants subsequently subjected to drought tolerance bioassay, showed that only the triple mutant plants (srk2d/e/i) died after seven days of water suppression, while all other mutants and wild type plants survived to maturity after rehydration. Obtained results suggest that the triple mutation completely blocks the ABA signaling and greatly affect the expression of various ABA/stress-responsive genes previously identified. Moreover, many osmotic stress marker genes (*KIN2, RD20,* and *COR15A RD29B*) are regulated by transcription factors that are controlled by protein kinases SRK2D/E/I that can act on ABA dependent and independent pathways in response to water stress.

Transgenic sugarcane plants overexpressing heterologous *P5CS* genes, responsible for the production of proline a protein commonly induced under stress conditions, revealed tolerance to severe water deficit, not as a mediator of osmotic adjustment, but as a component of the antioxidant defense system [89].

In the same way, after detecting the up-regulated expression of two maize putatives PIS in response to drought, one of them, the *ZmPIS* gene was over-expressed in tobacco plants showing to enhance drought tolerance, since it increased the membrane integrity and decreased the solute loss. The PIS gene is involved on the synthesis of phosphatidylinositol, an important lipid that functions as a key membrane constituent [90].

Remarkable results were observed with the *Arabidopsis* vacuolar pyrophosphatase gene (*AVP1*) over-expressed in cotton, which improved drought and salt tolerance in greenhouse conditions, and also increased fiber yield in dry land field conditions. Moreover, it was observed larger root systems and enhanced shoot biomass compared to controls when cultured under saline or reduced irrigation conditions [91].

4.2.2. Transcriptional factors genes

Transcription factors (TFs) have been extensively studied and have shown to be important in the regulation of stress tolerance in plants. The TFs are proteins that play a role in physiological and biological processes such as growth, development and responses to environmental stresses acting as key regulators involved in early stages of expression, gene regulation, signal transduction [92].

The TF *MYB15*, a member of the Arabidopsis *R2R3 MYB* family showed interesting results in studies carried out with Arabidopsis transgenic plants over-expressing the gene. It was found that the *MYB15* positively regulated tolerance to drought and salinity, inversely to what was observed in studies of freezing tolerance. Furthermore, the *MYB15* gene was found to be induced by treatment with ABA and salinity and drought conditions [83].

In *Oriza sativa* [93] reported that the gene *TSRF1* TF, a protein ERF (ethylene-responsive factor), when overexpressed increased drought and osmotic tolerance of transgenic rice plants, without affecting plant development. It also increased the sensitivity to ABA treatment, increased the content of proline and soluble sugars and the expression of genes related to responses to stress and photosynthesis. Curiously, in previous studies, the authors observed that *TSRF1* increased resistance to pathogens in tomato and tobacco plants, but reduced osmotic tolerance in tobacco.

The *sodERF3* another sugarcane TF, was also characterized, after *in silico* analyses of sugarcane cDNA sequence and the similarity of its 201 aa encoded proteins of the superfamily of ERF transcription factors was confirmed. A bioassay revealed that transgenic tobacco plants expressing the gene *sodERF3* showed to be tolerant to drought and osmotic stress. Furthermore, it was observed the gene induction in response to salt stress injuries and treatment with ABA [94].

Additionally, the soybean *GmERF3* TF from the *AP2/ERF* family was evaluated in transgenic tobacco plants and promoted tolerance to drought, salinity and disease. Furthermore, the gene expression was induced by salinity, desertification and treatments with salicylic acid (AS), ethylene (ET), JA and ABA. However, the cold stress did not affect gene expression. Thus, it is possible to assume that the *GmERF3* transcription factor plays a role in the responses to biotic and abiotic stresses [95].

It is evident the relevance of studying and elucidating the role of genes putatively related to water stress tolerance. In this context the molecular biology and the plant biotechnology comprise an efficient and helpful tool to achieve cultivars tolerant to environmental stresses that are gradually responsible for production losses all over the world.

4.2.3. Genetic modified crops using genetic engineering

Currently private companies have invested heavily in biotechnological programs for liberation of cultivars tolerant to insects, herbicides and drought. GMOs have been commercially cultivated since the 90's, tolerance to herbicides and insects are the main features of GM crops, including maize, soybean, cotton, canola, rice, tomato, etc. Some crops that has been transformed using genetic engineering technology to receive genes which metabolic function are related to water stress response are listed in Table 1. Genes involved in osmoprotection, ABA responsive elements and Transcription factors have been used to generate more resistant plants. Soybean, maize, rice, cotton and tomato are the most denoted transgenic crops.

Gene	Gene function	Metabolic Functions	Specie	References
adc	Arginine decarboxylase	Reduced chlorophyll loss under drought stress	*Oryza sativa*	[96]
Adc	Polyamine synthesis	Drought resistance	*Oryza sativa*	[97]
badh-1	Betaine aldehyde dehydrogenase	Maintenance of osmotic potential	*Solanum lycopersicum*	[98]
badh-1	Betaine aldehyde dehydrogenase	Salinity tolerance	*Daucus carota*	[99]

Gene	Gene function	Metabolic Functions	Specie	References
betA	Choline dehydrogenase (glycinebetaine synthesis)	Drought resistance at seedling stage and high yield after drought	*Zea mays*	[100]
codA	Choline oxidase (glycine betaine synthesis)	Tolerance to stress induced photo inhibition	*Brassica juncea*	[101]
codA	Choline oxidase (glycine betaine synthesis)	Increased tolerance to salinity and cold	*Oryza sativa*	[102]
codA	Choline oxidase (glycine betaine synthesis)	Recovery from a week long salt stress	*Oryza sativa*	[103]
cox	Choline oxidase (glycine betaine synthesis)	Salt and stress tolerance	*Oryza sativa*	[104]
gs2	Chloroplastic glutamine synthetase	Increased salinity resistance and chilling tolerance	*Oryza sativa*	[105]
mt1D and GutD	Mannitol-1-phosphate dehydrogenase & glucitol-6-phosphate dehydrogenase	High salt tolerance due to mannitol and glucitol accumulation	*Pinus taeda*	[106]
mt1D	Mannitol-1-phosphate dehydrogenase (mannitol synthesis)	Drought and salinity tolerance of calli and plants	*Triticum vulgare*	[107]
p5cs	Pyrroline carboxylate synthase (proline synthesis)	Osmotic adjustment and drought Resistance	*Citrus*	[108]
p5cs	Pyrroline carboxylate synthase (proline synthesis)	Salinity tolerance	*Solanum tuberosum*	[109]
p5cs	Pyrroline carboxylate synthase (proline synthesis)	Increased biomass production under drought and salinity stress	*Oryza sativa*	[110]
p5cs	Pyrroline carboxylate synthase (proline synthesis)	Reduced oxidative stress under osmotic stress	*Oryza sativa*	[111]
p5cs	Pyrroline carboxylate synthase (proline synthesis)	Drought resistance, high RWC, high proline	*Glycine max*	[112]
p5cs	Pyrroline carboxylate synthase (proline synthesis)	Drought resistance via antioxidant role of proline	*Saccharum spp.*	[89]

Gene	Gene function	Metabolic Functions	Specie	References
sst/fft	Fructan accumulation	Reduced proline accumulation at low water status	*Solanum tuberosum*	[113]
tpsp	Trehalose synthesis	Drought, salt and cold tolerance	*Oryza sativa*	[114]
tps1	Trehalose synthesis	Drought, salt and oxidative stress Tolerance	*Solanum lycopersicum*	[115]
hva1	Group 3 LEA protein gene	Delayed wilting under drought stress	*Avena sativa*	[116]
hva1	Group 3 LEA protein gene	Salinity tolerance in yield/ plant	*Avena sativa*	[117]
hva1	Group 3 LEA protein gene	Dehydration avoidance and cell membrane Stability	*Oryza saitva*	[118]
hva1	Group 3 LEA protein gene	Drought and salinity tolerance	*Oryza saitva*	[119]
hva1	Group 3 LEA protein gene	Increased biomass and WUE under stress	*Triticum vulgare*	[120]
hva1	Group 3 LEA protein gene	Improved plant water status and yield under field drought conditions	*Triticum vulgare*	[121]
OsLEA3-1	Lea protein	Drought resistance for yield in the field	*Oryza sativa*	[122]
rwc3	Aquaporin overexpression	Maintenance of leaf water potential and transpiration under 10 h PEG stress	*Oryza sativa*	[123]
atnhx1	Vacuolar Na^+/H^+ antiporter	Salt tolerance in photosynthesis and yield	*Gossypium hirsutum*	[124]
atnhx1	Vacuolar Na^+/H^+ antiporter	Salt tolerance, growth, fruit yield	*Solanum lycopersicum*	[125]
atnhx1	Vacuolar Na^+/H^+ antiporter	Salt tolerance for grain yield in the field	*Triticum vulgare*	[126]
hkt1	Potassium transporter	Salt tolerance in growth and improved K+/Na+ ratio	*Triticum vulgare*	[127]
abf3	Transcription factor	Drought resistance	*Oryza sativa*	[128]
adc	Arginine decarboxylase overexpression	Polyamine accumulation and salt resistance in biomass accumulation	*Oryza saitva*	[129]

Gene	Gene function	Metabolic Functions	Specie	References
dreb1 or osdreb1	Transcription factor	Drought, salt and cold tolerance with reduced growth under non-stress	*Oryza sativa*	[130]
dreb1a	Transcription factor	Delayed wilting under drought stress	*Triticum vulgare*	[131]
sp12 and sp5	ABA overproduction	High water-use efficiency, low transpiration and greater root hydraulic conductance	*Solanum lycopersicum*	[132]
tos1	Increased ABA sensitivity	Hypersensitive to osmotic stress and exogenous ABA	*Solanum lycopersicum*	[133]
ZmACS6	Ethylene synthesis	Non-functional mutant expressed drought induced Senescence	*Zea mays*	[134]
spcp2	putative papain-like cysteine protease	related to protein degradation for nutrient remobilization during leaf senescence	*Ipoema batatas*	[85]
srk2d, srk2e, srk2i	ABA signaling (ABA responsive element)	ABRE protein phosphorylation	*Arabidopsis thaliana*	[88]
zmpis	Phosphatidylinositol synthesis	Membrane protection	*Nicotiana tabacum*	[90]
avp1	Vacuolar pyrophosphatase gene	Proton pump activity	*Gossypium hirsutum*	[91]
myb15	Transcription factor (R2R3 MYB family member)	ABA responses, salinity and drought conditions tolerance	*Arabidopsis thaliana*	[83]
tsrf1	Protein ERF (ethylene-responsive factor)	ABA responses , accumulation of proline and soluble sugars content, induction of genes related to responses to stress and photosynthesis	*Oryza sativa*	[93]
soderf3	Sugarcane ERF transcription factors	Drought, osmotic stress, salt stress injuries and treatment with ABA	*Nicotiana tabacum*	[94]
ap2/erf	Soybean transcription factor family	Responses against biotic and abiotic stresses	*Nicotiana tabacum*	[95]

LEA Later embryogenesis abundant proteins

Table 1. Genes, gene function, metabolic function and transgenic.

http://dx.doi.org/10.5772/53877

5. Concluding remarks

The global warming is a reality that we have to face and in order to provide food to the growing population some actions have to be taken by government and researchers. The development of new cultivars more resistant to the environmental conditions, for every crop, must be a priority, in order to guarantee food demand security. To avoid yield reductions from floods, droughts and rising temperatures agribusiness will have to be reconsidered, investments have to be done, researchers will have to focus on ways to improve food quality, nutritional composition and increase yield using less land for farming. Crops will have to grow under a different scenario including less water and high temperature.

As it was discussed in this chapter changes in climate conditions will require several plant adaptations in order to minimize decreases in crop yield and to maintain food accessibility. Genetic breeding has been used over the last decades to improve yield and food quality. But how much can we still get from traditional plant breeding programs regarding to improve plants to face water scarcity? It is time to adopt new technologies as genetic engineering to help breeders in generating more adapted plants to survive water stress. For several years researchers have been spending time to understand how plants adapt to different situations, understanding the physiological parameters and their role in plants response to water stress specially hormones and transcriptional factors can help the development of new cultivars more resistant to stress conditions. All this knowledge allied to molecular biology techniques and genetic engineering can promote the development of transgenic plants with higher product quality, better storage conditions, easer processing, more efficient and more resistant to extreme conditions.

Nowadays, transgenic crops are cultivated all over the world, but there are some remains questions: How much are farmers dependent on biotechnology companies? Which economic and cultural losses transgenic cultures will bring about? These subjects are still extensively debated and researchers do not know for sure what is ahead. In the specific case of drought tolerance, much has been discussed about genetic engineering and experts consider the biotechnology relevant in developing higher genotypes.

Acknowledgements

The authors are grateful to: Fundação do Amparo à Pesquisa do Estado de São Paulo (FAPESP); Conselho Nacional de Desenvolvimento Cientifico e Tecnológico (CNPq) and Universidade de Ribeirão Preto (UNAERP) for the constant financial support and also to Ms Rosane Castro França for her precious help during this chapter organization.

Author details

Sonia Marli Zingaretti[1], Marielle Cascaes Inácio[2], Lívia de Matos Pereira[2], Tiago Antunes Paz[2] and Suzelei de Castro França[1]

1 Universidade de Ribeirão Preto, Brazil

2 Universidade Estadual Paulista, Brazil

References

[1] FAO-Food and Agriculture Organization of the United Nations. Crop Prospects and Food Situation 2012. Rome, Italy. http://www.fao.org/docrep/015/i2490e/i2490e00.htm (accessed 6 July 2012).

[2] Nelson GC., Rosegrant MW., Koo J, Robertson R, Sulser T, Zhu T, Ringler C, Msangi S, Palazzo A, Batka M, Magalhaes M, Valmonte-Santos R, Ewing M, LeeNelson DGC. Climate change: Impact on agriculture and costs of adaptation. In: Nelson GC., Rosegrant MW., Koo J, Robertson R, Sulser T, Zhu T, Ringler C, Msangi S, Palazzo A, Batka M, Magalhaes M, Valmonte-Santos R, Ewing M, LeeNelson DGC. Washington: International Food Policy Research Institute; 2009. 30p.

[3] Ciscar JC, Iglesias A, Feyen L, Szabo L, Van Regemorter D, Amelung B, Nicholls R, Watkiss P, Christensen OB, Dankers R, Garrote L, Goodess C M, Hunt A, Moreno A, Richards J, Soria A. Physical and economic consequences of climate change in Europe. Proceedings of the National Academy of Sciences of the United States of America 2011;108(7)2378-2683.

[4] Frei C, Scholl R, Fukutome S, Schmidli J, Vidale PL. Future change of precipitation extremes in Europe: Intercomparison of scenarios from regional climate models. Journal of Geophysical Research-Atmospheres 2006;111(D6)1-22.

[5] Tezara W, Mitchell VJ, Driscoll SD, Lawlor DW. Water stress inhibits plant photosynthesis by decreasing coupling factor and ATP. Nature 1999;401(6756)914-917.

[6] Ribas-Carbo M, Taylor NL, Giles L, Busquets S, Finnegan PM, Day DA, Lambers H, Medrano H, Berry JA, Flexas J. Effects of water stress on respiration in soybean leaves. Plant Physiology 2005;139(1)466-473.

[7] Guerfel M, Baccouri O, Boujnah D, Chaibi W, Zarrouk M. Impacts of water stress on gas exchange, water relations, chlorophyll content and leaf structure in the two main Tunisian olive (*Olea europaea* L.) cultivars. Scientia Horticulturae 2009;119(3)257-263.

[8] Silva EN, Ribeiro RV, Ferreira-Silva SL, Viegas RA, Silveira JAG. Comparative effects of salinity and water stress on photosynthesis, water relations and growth of *Jatropha curcas* plants. Journal of Arid Environments 2010;74(10)1130-1137.

[9] Sharkey TD, Seemann JR. Mild water-stress effects on carbon-reduction-cycle intermediates, ribulose bisphosphate carboxylase activity, and spatial homogeneity of photosynthesis in intact leaves. Plant Physiology 1989;89(4)1060-1065.

[10] Vassey TL, Sharkey TD. Mild water-stress of *Phaseolus vulgaris* plants leads to reduced starch synthesis and extractable sucrose phosphate synthase activity. Plant Physiology 1989;89(4)1066-1070.

[11] Chaves MM, Flexas J, Pinheiro C. Photosynthesis under drought and salt stress: regulation mechanisms from whole plant to cell. Annals of Botany 2009;103(4)551-560.

[12] Flexas J, Baron M, Bota J, Ducruet J-M, Galle A, Galmes J, Jimenez M, Pou A, Ribas-Carbo M, Sajnani C, Tomas M, Medrano H. Photosynthesis limitations during water stress acclimation and recovery in the drought-adapted *Vitis* hybrid Richter-110 (V. berlandieri×V. rupestris). Journal of Experimental Botany 2009;60(8)2361-2377.

[13] Yu J, Chen L, Xu M, Huang B. Effects of elevated CO_2 on physiological responses of tall fescue to elevated temperature, drought stress, and the combined stresses. Crop Science 2012;52(4)1848-1858.

[14] Galmes J, Ribas-Carbo M, Medrano H, Flexas J. Rubisco activity in Mediterranean species is regulated by the chloroplastic CO_2 concentration under water stress. Journal of Experimental Botany 2011;62(2)653-665.

[15] Carmo-Silva AE, Gore MA, Andrade-Sanchez P, French AN, Hunsaker DJ, Salvucci ME. Decreased CO_2 availability and inactivation of Rubisco limit photosynthesis in cotton plants under heat and drought stress in the field. Environmental and Experimental Botany 2012;83(1)1-11.

[16] Rivero RM, Shulaev V, Blumwald E. Cytokinin-Dependent Photorespiration and the Protection of Photosynthesis during Water Deficit. Plant Physiology 2009;150(3)1530-1540.

[17] Zhang N, Zhao B, Zhang HJ, Weeda S, Yang C, Yang ZC, Ren S, Guo Y. Melatonin Promotes Water-Stress Tolerance, Lateral Root Formation, and Seed Germination in Cucumber (*Cucumis sativus* L.). Journal of Pineal Research 2012;1-9. http://onlinelibrary.wiley.com/doi/10.1111/j.1600-079X.2012.01015.x/pdf (accessed 16 July 2012).

[18] Sanchez-Rodriguez E, del Mar Rubio-Wilhelmi M, Blasco B, Leyva R, Romero L, Manuel Ruiz J. Antioxidant response resides in the shoot in reciprocal grafts of drought-tolerant and drought-sensitive cultivars in tomato under water stress. Plant Science 2012;188-189(1)89-96.

[19] Jaleel CA, Manivannan P, Wahid A, Farooq M, Al-Juburi HJ, Somasundaram R, Panneerselvam R. Drought stress in plants: a review on morphological characteristics and pigments composition. International Journal of Agriculture and Biology 2009;11(1)100-105.

[20] Sangtarash MH, Qaderi MM, Chinnappa CC, Reid DM. Differential sensitivity of canola (*Brassica napus*) seedlings to ultraviolet-B radiation, water stress and abscisic acid. Environmental and Experimental Botany 2009;66(2)212-219.

[21] Pinheiro C, Antonio C, Ortuno MF, Dobrev PI, Hartung W, Thomas-Oates J, Ricardo CP, Vanková R, Chaves M, Wilson JC. Initial water deficit effects on *Lupinus albus* photosynthetic performance, carbon metabolism, and hormonal balance: metabolic reorganization prior to early stress responses. Journal of Experimental Botany 2011;62(14)4965-4974.

[22] Foito A, Byrne SL, Shepherd T, Stewart D, Barth S. Transcriptional and metabolic profiles of *Lolium perenne* L. genotypes in response to a PEG-induced water stress. Plant Biotechnology Journal 2009;7(8)719-732.

[23] Smith DO. Abscisic acid: interactions with ethylene and reactive oxygen species in the regulation of root growth under water deficit. Msc thesis. University of Missouri; 2011.

[24] Nan R, Carman JG, Salisbury FB. Water stress, CO_2 and photoperiod influence hormone levels in wheat. Journal of Plant Physiology 2002;159(3)307-312.

[25] Wang C, Yang A, Yin H, Zhang J. Influence of water stress on endogenous hormone contents and cell damage of maize seedlings. Journal of Integrative Plant Biology 2008;50(4)427-434.

[26] Leach KA, Hejlek LG, Hearne LB, Nguyen HT, Sharp RE, Davis GL. Primary root elongation rate and abscisic acid levels of maize in response to water stress. Crop Science 2011;51(1)157-172.

[27] Acharya B, Assmann S. Hormone interactions in stomatal function. Plant Molecular Biology 2009; 69(4)451-62.

[28] Zhang K, Xia X, Zhang Y, Gan S-S. An ABA-regulated and Golgi-localized protein phosphatase controls water loss during leaf senescence in *Arabidopsis*. Plant Journal 2012;69(4)667-678.

[29] Yang JC, Zhang JH, Wang ZQ, Zhu QS, Liu LJ. Abscisic acid and cytokinins in the root exudates and leaves and their relationship to senescence and remobilization of carbon reserves in rice subjected to water stress during grain filling. Planta 2002;215(4)645-652.

[30] Yang JC, Zhang JH, Wang ZQ, Zhu QS, Liu LJ. Involvement of abscisic acid and cytokinins in the senescence and remobilization of carbon reserves in wheat subjected to water stress during grain filling. Plant Cell and Environment 2003;26(10)1621-1631.

[31] Spollen WG, LeNoble ME, Samuels TD, Bernstein N, Sharp RE. Abscisic acid accumulation maintains maize primary root elongation at low water potentials by restricting ethylene production. Plant Physiology 2000;122(3)967-976.

[32] Sharp RE. Interaction with ethylene: changing views on the role of abscisic acid in root and shoot growth responses to water stress. Plant Cell and Environment 2002;25(2)211-222.

[33] Yang JC, Zhang JH, Wang ZQ, Zhu QS, Wang W. Hormonal changes in the grains of rice subjected to water stress during grain filling. Plant Physiology 2001;127(1)315-323.

[34] Jiang MY, Zhang JH. Involvement of plasma-membrane NADPH oxidase in abscisic acid- and water stress-induced antioxidant defense in leaves of maize seedlings. Planta 2002;215(6)1022-1030.

[35] Jiang MY, Zhang JH. Water stress-induced abscisic acid accumulation triggers the increased generation of reactive oxygen species and up-regulates the activities of antioxidant enzymes in maize leaves. Journal of Experimental Botany 2002;53(379)2401-2410.

[36] Ye N, Zhu G, Liu Y, Li Y, Zhang J. ABA controls H_2O_2 accumulation through the induction of *OsCATB* in rice leaves under water stress. Plant and Cell Physiology 2011;52(4)689-698.

[37] Apelbaum A, Yang SF. Biosynthesis of stress ethylene induced by water deficit. Plant Physiology 1981;68(3)594-596.

[38] Vaadia Y. Plant hormones and water stress. Philosophical Transactions of the Royal Society of London Series B, Biological Sciences 1976;273(927)513-522.

[39] Merewitz EB, Gianfagna T, Huang B. Photosynthesis, water use, and root viability under water stress as affected by expression of *SAG12-ipt* controlling cytokinin synthesis in *Agrostis stolonifera*. Journal of Experimental Botany 2011;62(1)383-395.

[40] Tognetti VB, Van Aken O, Morreel K, Vandenbroucke K, van de Cotte B, De Clercq I, Chiwocha S, Fenske R, Prinsen E, Boerjan W, Genty B, Stubbs KA, Inzé D, Van Breusegem F. Perturbation of indole-3-butyric acid homeostasis by the UDP-Glucosyltransferase UGT74E2 modulates *Arabidopsis* architecture and water stress tolerance. Plant Cell 2010;22(8)2660-2679.

[41] De Ollas C, Hernando B, Arbona V, Gómez-Cadenas A. Jasmonic acid transient accumulation is needed for abscisic acid increase in citrus roots under drought stress conditions. Physiologia Plantarum 2012;1-11. http://onlinelibrary.wiley.com/doi/10.1111/j.1399-3054.2012.01659.x/pdf (accessed 3 August 2012).

[42] Maes WH, Achten WMJ, Reubens B, Raes D, Samson R, Muys B. Plant-water relationships and growth strategies of *Jatropha curcas* L. seedlings under different levels of drought stress. Journal of Arid Environments 2009;73(10)877-884.

[43] O'Toole J, Cruz RT. Response of leaf water potential, stomatal resistance, and leaf rolling to water stress. Plant Physiology 1980;65(3)428-432.

[44] Ehlert C, Plassard C, Cookson SJ, Tardieu F, Simonneau T. Do pH changes in the leaf apoplast contribute to rapid inhibition of leaf elongation rate by water stress? Com-

parison of stress responses induced by polyethylene glycol and down-regulation of root hydraulic conductivity. Plant Cell and Environment 2011;34(8)1258-1266.

[45] Brodribb TJ, Holbrook NM. Water stress deforms tracheids peripheral to the leaf vein of a tropical conifer. Plant Physiology 2005;137(3)1139-1146.

[46] Schuppler U, He PH, John PCL, Munns R. Effect of water stress on cell division and cell-division-cycle 2-like cell-cycle kinase activity in wheat leaves. Plant Physiology 1998;117(2)667-678.

[47] Ackerson RC, Hebert RR. Osmoregulation in cotton in response to water stress: I. Alterations in photosynthesis, leaf conductance, translocation, and ultrastructure. Plant Physiology 1981;67(3): 484-488.

[48] Nayyar H, Gupta D. Differential sensitivity of C3 and C4 plants to water deficit stress: Association with oxidative stress and antioxidants. Environmental and Experimental Botany 2006;58(1-3)106-113.

[49] Ghannoum O. C4 photosynthesis and water stress. Annals of Botany 2009;103(1)635-644.

[50] Alfonso SU, Brüggemann W. Photosynthetic responses of a C3 and three C4 species of the genus *Panicum* (s.l.) with different metabolic subtypes to drought stress. Photosynthesis Research 2012;1-17. http://www.springerlink.com/content/202130j47376816n/fulltext.pdf (accessed 3 August 2012).

[51] Chen JW, Bai KD, Cao KF. Inhibition of monoterpene biosynthesis accelerates oxidative stress and leads to enhancement of antioxidant defenses in leaves of rubber tree (*Hevea brasiliensis*). Acta Physiologiae Plantarum 2009;31(1)95-101.

[52] Turtola S, Manninen AM, Rikala R, Kainulainen P. Drought stress alters the concentration of wood terpenoids in Scots pine and Norway spruce seedlings. Journal of Chemical Ecology 2003;29(9)1981-1995.

[53] Chen Y, Guo Q, Liu L, Liao L, Zhu Z. Influence of fertilization and drought stress on the growth and production of secondary metabolites in *Prunella vulgaris* L. Journal of Medicinal Plants Research 2011;5(9)1749-1755.

[54] Hura T, Hura K, Grzesiak S. Possible contribution of cell-wall-bound ferulic acid in drought resistance and recovery in triticale seedlings. Journal of Plant Physiology 2009;166(16)1720-1733.

[55] Azhar N, Hussain B, Ashraf MY, Abbasi KY. Water stress mediated changes in growth, physiology and. secondary metabolites of desi ajwain (*Trachyspermum ammi* L.). Pakistan Journal of Botany 2011;43(SI)15-19.

[56] Amooaghaie R. Role of polyamines in the tolerance of soybean to water deficit stress. World Academy of Science, Engineering and Technology 2011;(56)498-502.

[57] Nacif de Abreu I, Mazzafera P. Effect of water and temperature stress on the content of active constituents of *Hypericum brasiliense* Choisy. Plant Physiology and Biochemistry 2005;43(3)241-48.

[58] Velloso MAL, Abreu IN, Mazzafera P. Indução de metabólitos secundários em plântulas de *Hypericum brasiliense* Choisy crescendo *in vitro*. Acta Amazonica 2009;39(2)267-72.

[59] Marchese JA, Ferreira JFS, Rehder VLG, Rodrigues O. Water deficit effect on the accumulation of biomass and artemisinin in annual worm wood (*Artemisia annua* L., Asteraceae). Brazilian Journal of Plant Physiology 2010;22(1)1-9.

[60] Jaleel CA, Manivannan P, Sankar B, Kishorekumar A, Gopi R, Somasundaram R, Panneerselvam R. Induction of drought stress tolerance by ketoconazole in *Catharanthus roseus* is mediated by enhanced antioxidant potentials and secondary metabolite accumulation. Colloids and Surfaces B-Biointerfaces 2007;60(2)201-206.

[61] Bosswell MJ. Plant Oils: Wealth, health, energy and environment. In: Proceedings International Conference of Renewable Energy Technology for Rural Development; 2003.

[62] Hanna M, Isom L, Campbell J. Biodiesel: current perspectives and future. Journal of Scientific and Industrial Research 2005;64(11)854-857.

[63] Pousa GPAG, Santos ALF, Suarez PAZ. History and policy of biodiesel in Brazil. Energy Policy 2007;35(11)5393-5398.

[64] Martini LH, Jung F, Soares FA, Rotta LN, Vendite DA, dos Santos Frizzo ME, Yunes RA, Calixto JB, Wofchuk S, Souza DO. Naturally occurring compounds affect glutamatergic neurotransmission in rat brain. Neurochemical Research 2007;32(11)1950-1956.

[65] Theoduloz C, Rodriguez JA, Pertino M, Schmeda-Hirschmann G. Antiproliferative activity of the diterpenes jatrophone and jatropholone and their derivatives. Planta Medica 2009;75(14)1520-1522.

[66] Koziol M. Chemical composition and nutritional evaluation of quinoa (*Chenopodium quinoa* Willd.). Journal of food composition and analysis: an official publication of the United Nations University, International Network of Food Data Systems 1992;5(1)35-68.

[67] Jacobsen SE. The worldwide potential for quinoa (*Chenopodium quinoa* Willd.). Food Reviews International 2003;19(1-2)167-177.

[68] Gorinstein S, Vargas OJM, Jaramillo NO, Salas IA, Ayala ALM, Arancibia-Avila P, Toledo F, Katrich E, Trakhtenberg S. The total polyphenols and the antioxidant potentials of some selected cereals and pseudocereals. European Food Research and Technology 2007;225(3-4)321-328.

[69] Nsimba RY, Kikuzaki H, Konishi Y. Antioxidant activity of various extracts and fractions of *Chenopodium quinoa* and *Amaranthus* spp. seeds. Food Chemistry 2008;106(2)760-766.

[70] Repo-Carrasco-Valencia RAM, Serna AL. Quinoa (*Chenopodium quinoa* Willd.) as a source of dietary fiber and other functional components. Ciencia e Tecnologia de Alimentos 2011;31(1)225-230.

[71] Yildiz-Aktas L, Dagnon S, Gurel A, Gesheva E, Edreva A. Drought tolerance in cotton: involvement of non-enzymatic ROS-Scavenging Compounds. Journal of Agronomy and Crop Science 2009;195(4)247–253.

[72] Witcombe JR, Hollington PA, Howarth CJ, Reader S, Steele KA. Breeding for abiotic stresses for sustainable agriculture. Philosophical Transactions of the Royal Society B-Biological Sciences 2008;363(1492)703–716.

[73] Peleg Z, Walia H, E. B. Integrating genomics and genetics to accelerate development of drought and salinity tolerant crops. In: Altman A, Hasegawa PM. (ed.) Plant Biotechnology and Agriculture – Prospects for the 21st century. London: Elsevier; 2012. p 271-281.

[74] FAO- Food and Agriculture Organization of the United Nations. Biotechnologies for Agriculture Development. Current status and options for crop biotechnology in developing countries. Rome, Italy. http://www.fao.org/docrep/014/i2300e/i2300e00.htm (accessed 1 July 2012).

[75] Bray EA. Molecular responses to water-deficit. Plant Physiology 1993;103(4)1035-1040.

[76] Rodrigues F, Da Graça J, De Laia M, Nhani-Jr A, Galbiati J, Ferro MIT, Ferro, J.; Zingaretti, S. Sugarcane genes differentially expressed during water deficit. Biologia Plantarum 2011;55(1)43-53.

[77] Bray EA. Genes commonly regulated by water-deficit stress in Arabidopsis thaliana. Journal of Experimental Botany 2004;55(407)2331–2341.

[78] Bray EA. Plant responses to water deficit. Trends in Plant Science 1997;2(2)1035-1040.

[79] Shao H-B, Guo Q-J, Chu L-Y, Zhao X-N, Su Z-L, Hu Y-C, Cheng JF. Understanding molecular mechanism of higher plant plasticity under abiotic stress. Colloids and Surfaces B-Biointerfaces 2007;54(1)37-45.

[80] Shinozaki K, Yamaguchi-Shinozaki K. Molecular responses to drought and cold stress. Current Opinion in Biotechnology 1996;7(2)161-167.

[81] Rodrigues FA, de Laia ML, Zingaretti SM. Analysis of gene expression profiles under water stress in tolerant and sensitive sugarcane plants. Plant Science 2009;176(2)286–302.

[82] Iskandar HM, Casu RE, Fletcher AT, Schmidt S, Xu J, Maclean DJ, Manners JM, Bonnett GD. Identification of drought-response genes and a study of their expression

during sucrose accumulation and water deficit in sugarcane culms. BMC Plant Biology 2011;11-12 http://www.biomedcentral.com(accessed 4 June 2012).

[83] Ding Z, Li S, An X, Liu X, Qin H, Wang D. Transgenic expression of *MYB15* confers enhanced sensitivity to abscisic acid and improved drought tolerance in Arabidopsis thaliana. Journal of Genetics and Genomics 2009;36(1)17-29.

[84] Wang WX, Vinocur B, Altman A. Plant responses to drought, salinity and extreme temperatures: towards genetic engineering for stress tolerance. Planta 2003;218(1)2181-2184.

[85] Chen H-J, Su C-T, Lin C-H, Huang G-J, Lin Y-H. Expression of sweet potato cysteine protease SPCP2 altered developmental characteristics and stress responses in transgenic Arabidopsis plants. Journal of Plant Physiology 2010;167(10)838–847.

[86] Nogueira FTS, Schlogl PS, Camargo SR, Fernandez JH, De Rosa VE, Pompermayer P, Arruda P. *SsNAC23*, a member of the NAC domain protein family, is associated with cold, herbivory and water stress in sugarcane. Plant Science 2005;169(1)93–106.

[87] Agarwal PK, Agarwal P, Reddy MK, Sopory SK. Role of DREB transcription factors in abiotic and biotic stress tolerance in plants. Plant Cell Reports 2006;25(12)1263-1274.

[88] Fujita Y, Nakashima K, Yoshida T, Katagiri T, Kidokoro S, Kanamori N, Umezawa T, Fujita M, Maruyama K, Ishiyama K, Kobayashi M, Nakasone S, Yamada K, Ito T, Shinozaki K, Yamaguchi-Shinozaki K. Three SnRK2 protein kinases are the main positive regulators of abscisic acid signaling in response to water stress in *Arabidopsis*. Plant and Cell Physiology 2009;50(12)2123–2132.

[89] Molinari HBC, Marur CJ, Daros E, de Campos MKF, de Carvalho JFRP, Bespalhok Filho JC, Pereira LFP, Vieira LGE. Evaluation of the stress-inducible production of proline in transgenic sugarcane (*Saccharum* spp.): osmotic adjustment, chlorophyll fluorescence and oxidative stress. Physiologia Plantarum 2007;130(2)218–229.

[90] Zhai S-M, Gao Q, Xue H-W, Sui Z-H, Yue G-D, Yang A-F, Zhang J-R. Overexpression of the phosphatidylinositol synthase gene from *Zea mays* in tobacco plants alters the membrane lipids composition and improves drought stress tolerance. Planta 2012;235(1)69–84.

[91] Zhang H, Shen G, Kuppu S, Gaxiola R, Payton P. Creating drought-and salt-tolerant cotton by overexpressing a vacuolar pyrophosphatase gene. Plant signaling & behavior 2011;6(6)861-863.

[92] Yu S, Liao F, Wang F, Wen W, Li J, Mei H, Luo L. Identification of Rice Transcription Factors Associated with Drought Tolerance Using the Ecotilling Method. Plos One 2012;7(2)1-9.

[93] Quan R, Hu S, Zhang Z, Zhang H, Zhang Z, Huang R. Overexpression of an ERF transcription factor *TSRF1* improves rice drought tolerance. Plant Biotechnology Journal 2010;8(4)476–488.

[94] Trujillo LE, Sotolongo M, Menendez C, Ochogavia ME, Coll Y, Hernandez I, Borrás-Hidalgo O, Thomma BPHJ, Vera P, Hernández L. *SodERF3*, a novel sugarcane ethylene responsive factor (ERF), enhances salt and drought tolerance when overexpressed in tobacco plants. Plant and Cell Physiology 2008;49(4)512–525.

[95] Zhang G, Chen M, Li L, Xu Z, Chen X, Guo J, Ma Y. Overexpression of the soybean *GmERF3* gene, an *AP2/ERF* type transcription factor for increased tolerances to salt, drought, and diseases in transgenic tobacco. Journal of Experimental Botany 2009;60(13)3781–3796.

[96] Capell T, Escobar C, Liu H, Burtin D, Lepri O, Christou P. Over-expression of the oat arginine decarboxylase cDNA in transgenic rice (*Oryza sativa* L.) affects normal development patterns in vitro and results in putrescine accumulation in transgenic plants. Theoretical and Applied Genetics 1998;97(1-2)246-254.

[97] Capell T, Bassie L, Christou P. Modulation of the polyamine biosynthetic pathway in transgenic rice confers tolerance to drought stress. Proceedings of the National Academy of Sciences of the United States of America 2004;101(26)9909–9914.

[98] Moghaieb REA, Tanaka N, Saneoka H, Hussein HA, Yousef SS, Ewada MAF, Aly MAM, Fujita K. Expression of betaine aldehyde dehydrogenase gene in transgenic tomato hairy roots leads to the accumulation of glycine betaine and contributes to the maintenance of the osmotic potential under salt stress. Soil Science and Plant Nutrition 2000;46(4)873-883.

[99] Kumar S, Dhingra A, Daniell H. Plastid-expressed betaine aldehyde dehydrogenase gene in carrot cultured cells, roots, and leaves confers enhanced salt tolerance. Plant Physiology 2004;136(1)2843-2854.

[100] Quan R, Shang M, Zhang H, Zhao Y, Zhang J. Engineering of enhanced glycine betaine synthesis improves drought tolerance in maize. Plant Biotechnol Journal 2004;2(6)477-486.

[101] Prasad K, Saradhi PP. Enhanced tolerance to photoinhibition in transgenic plants through targeting of glycinebetaine biosynthesis into the chloroplasts. Plant Science 2004;166(5)1197-1212.

[102] Sakamoto A, Murata A, Murata N. Metabolic engineering of rice leading to biosynthesis of glycinebetaine and tolerance to salt and cold. Plant Molecular Biology 1998;38(6)1011-1019.

[103] Mohanty A, Kathuria H, Ferjani A, Sakamoto A, Mohanty P, Murata N, Tyagi AK. Transgenics of an elite indica rice variety Pusa Basmati 1 harbouring the codA gene are highly tolerant to salt stress. Theoretical and Applied Genetics 2002;106(1)51-57.

[104] Su J, Hirji R, Zhang L, He CK, Selvaraj G, Wu R. Evaluation of the stress-inducible production of choline oxidase in transgenic rice as a strategy for producing the stress-protectant glycine betaine. Journal of Experimental Botany 2006;57(5)1129-1135.

[105] Hoshida H, Tanaka Y, Hibino T, Hayashi Y, Tanaka A, Takabe T. Enhanced tolerance to salt stress in transgenic rice that overexpresses chloroplast glutamine synthetase. Plant Molecular Biology 2000;43(1)103-111.

[106] Tang W, Peng XX, Newton RJ. Enhanced tolerance to salt stress in transgenic loblolly pine simultaneously expressing two genes encoding mannitol-1-phosphate dehydrogenase and glucitol-6-phosphate dehydrogenase. Plant Physiology and Biochemistry 2005;43(2)139-146.

[107] Abebe T, Guenzi AC, Martin B, Cushman JC. Tolerance of mannitol-accumulating transgenic wheat to water stress and salinity. Plant Physiology 2003;131(4)1748-1755.

[108] Molinari HBC, Marur CJ, Kobayashi AK, Pileggi M, Júnior RPL, Pereira LFP, Vieira LGE. Osmotic adjustment in transgenic citrus root stock Carrizo citrange (*Citrus sinensis* Osb. x *Poncirus trifoliata* L. Raf.) overproducing proline. Plant Science 2004;167(6)1375-81.

[109] Hmida-Sayari A, Gargouri-Bouzid R, Bidani A, Jaoua L, Savoure A, Jaoua S. Overexpression of Delta(1)-pyrroline-5-carboxylate synthetase increases proline production and confers salt tolerance in transgenic potato plants. Plant Science 2005;169(4)746-752.

[110] Zhu B, Su J, Chang M, Verma DPS, Fan YL, Wu R. Overexpression of a [delta] 1-pyrroline-5-carboxylate synthetase gene and analysis of tolerance to water-and salt-stress in transgenic rice. Plant Science 1998;139(1)41-48.

[111] Hong ZL, Lakkineni K, Zhang ZM, Verma DPS. Removal of feedback inhibition of Delta (1)-pyrroline-5-carboxylate synthetase results in increased proline accumulation and protection of plants from osmotic stress. Plant Physiology 2000; 122(4)1129-1136.

[112] De Ronde J, Cress W, Krüger G, Strasser R, Van Staden J. Photosynthetic response of transgenic soybean plants, containing an Arabidopsis *P5CR* gene, during heat and drought stress. Journal of Plant Physiology 2004;161(11)1211-24.

[113] Knipp G, Honermeier B. Effect of water stress on proline accumulation of genetically modified potatoes (*Solanum tuberosum* L.) generating fructans. Journal of Plant Physiology 2006;163(4)392-397.

[114] Jang IC, Oh SJ, Seo JS, Choi WB, Song SI, Kim CH, et al. Expression of a bifunctional fusion of the Escherichia coli genes for trehalose-6-phosphate synthase and trehalose-6-phosphate phosphatase in transgenic rice plants increases trehalose accumulation and abiotic stress tolerance without stunting growth. Plant Physiology 2003;131(2)516-524.

[115] Cortina C, Culianez-Macia FA. Tomato abiotic stress enhanced tolerance by trehalose biosynthesis. Plant Science 2005;169(1)75–82.

[116] Maqbool B, Zhong H, El-Maghraby Y, Ahmad A, Chai B, Wang W, Sabzikar R, Sticklen B. Competence of oat (*Avena sativa* L.) shoot apical meristems for integrative

transformation, inherited expression, and osmotic tolerance of transgenic lines containing hva1. Theoretical and Applied Genetics 2002;105(2-3)201-208.

[117] Oraby HF, Ransom CB, Kravchenko AN, Sticklen MB. Barley *HVA1* gene confers salt tolerance in R3 transgenic oat. Crop Science 2005;45(6)2218-2227.

[118] Babu RC, Zhang JX, Blum A, Ho THD, Wu R, Nguyen HT. *HVA1*, a LEA gene from barley confers dehydration tolerance in transgenic rice (Oryza sativa L.) via cell membrane protection. Plant Science 2004;166(4)855-862.

[119] Rohila JS, Jain RK, Wu R. Genetic improvement of Basmati rice for salt and drought tolerance by regulated expression of a barley Hva1 cDNA. Plant Science 2002;163(3)525-532.

[120] Sivamani E, Bahieldin A, Wraith JM, Al-Niemi T, Dyer WE, Ho THD, Qu RD. Improved biomass productivity and water use efficiency under water deficit conditions in transgenic wheat constitutively expressing the barley *HVA1* gene. Plant Science 2000;155(1)1-9.

[121] Bahieldin A, Mahfouz HT, Eissa HF, Saleh OM, Ramadan AM, Ahmed IA, Dyer WE, El-Itriby HA, Madkour MA. Field evaluation of transgenic wheat plants stably expressing the *HVA1* gene for drought tolerance. Physiologia Plantarum 2005;123(4)421-427.

[122] Xiao B, Huang Y, Tang N, Xiong L. Over-expression of a LEA gene in rice improves drought resistance under the field conditions. Theoretical and Applied Genetics 2007;115(1)35-46.

[123] Lian HL, Yu X, Ye Q, Ding XS, Kitagawa Y, Kwak SS, Su W-A, Tang Z-C. The role of aquaporin *RWC3* in drought avoidance in rice. Plant and Cell Physiology 2004;45(4)481-489.

[124] He C, Yan J, Shen G, Fu L, Holaday AS, Auld D, Blumwald E, Zhang H. Expression of an *Arabidopsis* vacuolar sodium/proton antiporter gene in cotton improves photosynthetic performance under salt conditions and increases fiber yield in the field. Plant and cell physiology 2005;46(11)1848-1854.

[125] Apse MP, Aharon GS, Snedden WA, Blumwald E. Salt tolerance conferred by overexpression of a vacuolar Na+/H+ antiport in *Arabidopsis*. Science 1999;285(5431)1256–1258.

[126] Xue ZY, Zhi DY, Xue GP, Zhang H, Zhao YX, Xia GM. Enhanced salt tolerance of transgenic wheat (*Triticum aestivum* L.) expressing a vacuolar Na+/H+ antiporter gene with improved grain yields in saline soils in the field and a reduced level of leaf Na+. Plant Science 2004;167(4)849-859.

[127] Laurie S, Feeney KA, Maathuis FJM, Heard PJ, Brown SJ, Leigh RA. A role for *HKT1* in sodium uptake by wheat roots. Plant Journal 2002;32(2)139-149.

http://dx.doi.org/10.5772/53877

[128] Oh SJ, Song SI, Kim YS, Jang HJ, Kim SY, Kim M, Nahm BH, Ju- Kim K. *Arabidopsis CBF3/DREB1A* and *ABF3* in transgenic rice increased tolerance to abiotic stress without stunting growth. Plant Physiology 2005;138(1)341-351.

[129] Roy M, Wu R. Arginine decarboxylase transgene expression and analysis of environmental stress tolerance in transgenic rice. Plant Science 2001;160(5)869-875.

[130] Ito Y, Katsura K, Maruyama K, Taji T, Kobayashi M, Seki M, Shinozaki K, Kazuko Y-S. Functional analysis of rice DREB1/CBF-type transcription factors involved in cold-responsive gene expression in transgenic rice. Plant and Cell Physiology 2006;47(1)141-153.

[131] Pellegrineschi A, Reynolds M, Pacheco M, Brito RM, Almeraya R, Yamaguchi-Shinozaki K, Hoisington D. Stress-induced expression in wheat of the *Arabidopsis thaliana DREB1A* gene delays water stress symptoms under greenhouse conditions. Genome 2004;47(3)493-500.

[132] Thompson AJ, Andrews J, Mulholland BJ, McKee JMT, Hilton HW, Horridge JS, Farquhar GD, Smeeton RC, Smillie IRA, Black CR, Taylor IB. Overproduction of abscisic acid in tomato increases transpiration efficiency and root hydraulic conductivity and influences leaf expansion. Plant Physiology 2007;143(4)1905-1917.

[133] Borsani O, Cuartero J, Valpuesta V, Botella MA. Tomato tos1 mutation identifies a gene essential for osmotic tolerance and abscisic acid sensitivity. Plant Journal 2002;32(6)905-914.

[134] Young TE, Meeley RB, Gallie DR. ACC synthase expression regulates leaf performance and drought tolerance in maize. Plant Journal 2004;40(5):813-825.